Visualization in Landscape and Environmental Planning

Why is visualization important? How is it done? What are the ethical issues? What are the major applications? Where are we going? In this book many of the world's leading researchers in the development and application of visualization for landscape and environmental management answer these questions. Environmental visualization involves technology and processes designed to provide a realistic view of environmental futures to support public understanding and debate on landscape issues.

The book begins with an introduction to the field of environmental visualization through its development in pre-history to the explosion of opportunity brought by computer developments in recent decades. The authors then consider the realistic simulation of the environment in a broader visualization context. The resources available for building landscape models are explored. The technology of visualization is a rich topic and here current developments in hardware, software and display technology are reviewed by leading developers. Important issues related to the validity of visual simulations and the ethics of their application are raised. Many authors have contributed state-of-the-art applications in forestry, agriculture, mining, energy and urban design. The volume finishes with an overview of prospects for the future with an emphasis on the potential of augmented realities and new approaches to public participation.

Ian D. Bishop is a professor of Geomatics at the University of Melbourne, Australia. He has researched and worked in the fields of environmental visualization, land use modelling, landscape assessment, and geographic information systems (GIS). In his current research he remains committed to the potential of GIS and visualization technology to improve the management of natural resources and to contribute to public participation in planning and design issues.

Eckart Lange is a professor in the Department of Landscape at the University of Sheffield, UK. Educated as a landscape planner and landscape architect at the TU Berlin, Heriot-Watt-University Edinburgh and Harvard University, for many years he was head of the landscape research group at the Swiss Federal Institute of Technology (ETH Zürich). His research and practice is dedicated to landscape visualization being of major instrumental importance for the design, planning and sustainable development of landscape.

Visualization in Landscape and Environmental Planning

Technology and Applications

Edited by Ian D. Bishop and Eckart Lange

Taylor & Francis
Taylor & Francis Group
LONDON AND NEW YORK

First published 2005 by Taylor & Francis
2 Park Square, Milton Park, Abingdon, Oxon OX14 4RN

Simultaneously published in the USA and Canada by Taylor & Francis
270 Madison Ave, New York, NY 10016

Taylor & Francis is an imprint of the Taylor & Francis Group

© 2005 Ian D. Bishop and Eckart Lange, editorial and selection; individual chapters, the contributors

Designed and typeset in Sabon by Alex Lazarou, Surbiton, Surrey
Printed and bound in Great Britain by St Edmunsbury Press, Bury St Edmunds, Suffolk

All rights reserved. No part of this book may be reprinted or reproduced or utilised in any form or by any electronic, mechanical, or other means, now known or hereafter invented, including photocopying and recording, or in any information storage or retrieval system, without permission in writing from the publishers.

British Library Cataloguing in Publication Data
A catalogue record for this book is available from the British Library

Library of Congress Cataloging in Publication Data
Visualization in landscape and environmental planning: technology and applications/ edited by Eckart Lange and Ian D. Bishop.
    p. cm.
    Includes bibliographical references and index.
    ISBN 0-415-30510-1 (hardcover: alk. paper) — ISBN 0-203-53200-7 (ebook)
    1. Landscape architecture — Computer-aided design. 2. Landuse — Planning — Computer-aided design. 3. Visualization. I. Lange, Eckart. II. Bishop, Ian D.

SB475.9.D37V57 2005
712'.0285—dc22

2004016566

ISBN 0-415-30510-1

CONTENTS

Preface — ix

Foreword — xi
Stephen M. Ervin

Contributors — xv

Acknowledgements — xxi

Part 1   Understanding visualization — 1

1 Communication, perception and visualization — 3
   Eckart Lange and Ian D. Bishop

2 Visualization classified — 23
   Ian D. Bishop and Eckart Lange

3 Data sources for three-dimensional models — 35
   Ben Discoe

4 Visualization technology — 51

   Introduction — 51
   Ian D. Bishop and Eckart Lange

   Efficient modelling and rendering of landscapes — 56
   Oliver Deussen, Carsten Colditz, Liviu Coconu and Hans-Christian Hege

	Using games software for interactive landscape visualization *Adrian Herwig, Einar Kretzler and Philip Paar*	62
	Presentation style and technology *Ian D. Bishop and Eckart Lange*	68
5	**Validity, reliability and ethics in visualization** *Stephen R.J. Sheppard*	79

	Part 2 Applications	99
6	**Applications in the forest landscape**	101
	Introduction *Duncan Cavens*	101
	'Calibrating' images to more accurately represent future landscape conditions in forestry *Brian Orland*	104
	Studying the acceptability of forest management practices using visual simulation of forest regrowth *Ian D. Bishop, Rebecca Ford, Daniel Loiterton and Kathryn Williams*	112
	Planning, communicating, designing and decision making for large scale landscapes *Jon Salter, Stephen R.J. Sheppard, Duncan Cavens and Michael Meitner*	120
	The role of landscape simulators in forestry: a Finnish perspective *Liisa Tyrväinen and Jori Uusitalo*	125
7	**Applications in the agricultural landscape**	133
	Introduction *Ian D. Bishop*	133
	Designing, visualizing and evaluating sustainable agricultural landscapes *Andrew Lovett*	136
	Helping rural communities envision their future *Christian Stock and Ian D. Bishop*	145
	Lenné3D – walk-through visualization of planned landscapes *Philip Paar and Jörg Rekittke*	152

8	**Applications in energy, industry and infrastructure**	163
	Introduction *Ian D. Bishop*	163
	Visualizing scenic resource impacts: proposed surface mining and solid waste sanitary landfill *John C. Ellsworth, Abraham N. Medina and Issa A. Hamud*	166
	The provision of visualization tools for engaging public and professional audiences *David R. Miller, Jane G. Morrice and Alastor Coleby*	175
	The visualization of windfarms *John F. Benson*	184
9	**Applications in the urban landscape**	193
	Introduction *Eckart Lange*	193
	Future scenarios of peri-urban green space *Eckart Lange and Sigrid Hehl-Lange*	195
	Negotiating public view protection and high density in urban design *John W. Danahy*	203
	Combining visualization with choice experimentation in the built environment *Richard Laing, Anne-Marie Davies and Stephen Scott*	212
	Part 3 Prospects	221
10	**Visualization prospects**	223
	Introduction *Ian D. Bishop and Eckart Lange*	223
	Compositing of computer graphics with landscape video sequences *Eihachiro Nakamae, Xueying Qin and Katsumi Tadamura*	226
	Future use of augmented reality for environmental and landscape planners *Wayne Piekarski and Bruce H. Thomas*	234
	Visualization and participatory decision making *Nathan H. Perkins and Steve Barnhart*	241

Visualization in support of public participation *Michael Kwartler*	251
Trends, challenges and a glimpse of the future *Ian D. Bishop and Eckart Lange*	261
References	267
Index	293

PREFACE

In the past few years, significant advances in computer graphics have created exciting new options for visualizing our environment in three or four dimensions (with animation over time or movement through space). These advances are of major interest to the landscape and environmental professions. Recent important developments have included highly realistic representation of vegetation, efficient display of terrain, and automatic generation of landscape models and images from GIS-based data. The technology that exists today enables us to render visually stunning and richly detailed visual simulations of natural and urban environments.

The heart of this book is the application case studies (Chapters 6–9) that present the state-of-the-art in environmental visualization. We have divided these, to the degree possible, by the activity or land cover type (forestry, agriculture, mining and industry and urban) to which they apply, recognizing that substantial levels of overlap are inevitable. Preceding the applications chapters are:

- a brief history of developments in landscape visualization;
- an attempt to classify the range of activities which are broadly referred to as visualization and to distinguish what we are calling landscape or environmental visualization from the remainder of the visualization world;
- a discussion of digital data sources for environmental visualization, recognizing that increasing volumes of digital data are available worldwide;
- a review of the salient characteristics of hardware and software to assist a newcomer to begin working in visualization;
- to help that newcomer, and the more experienced amongst us, to do the job properly there is a discussion of validity issues and some guidelines for appropriate and effective visualization.

After the case studies we look into the future and consider the developments (both technological and cultural) that appear to be just around the corner.

From this description it should be clear that this is not a 'how to' book in the sense of dwelling on computer algorithms or particular software products. Many procedures for simulating particular aspects of the landscape emerge in the case studies but this is not a comprehensive review. For more detail on the options for modelling a tree, a forest, a plume of smoke or other major environmental features we refer you to Ervin and Hasbrouck (2001).

Ian Bishop, Melbourne
Eckart Lange, Zürich and Sheffield

FOREWORD

In the spirit of this book about visualization, perhaps a single image should be here in place of these thousand or so words, demonstrating the truth of that old adage about their equivalency. But we all know that the truth of that saying is dependent upon context and application, for we understand that words and images serve different, albeit often mutually reinforcing, purposes. As this well-illustrated volume shows, both words and images are profitably used, often together, to engage the senses – and thus the intellect – in the processes of communication. One interpretation of this book about visualization is that it's largely about communication.

Because the editors, Ian Bishop and Eckart Lange, and many of their intended audience, are planners and designers (even if they don't have those words in their titles or job descriptions) this book is also about planning and design; specifically, it's about visual communications in environmental planning and design. A distinction is made in the second chapter between 'communicating' and 'discovering knowledge', and the different roles of visualization in these respective activities; but it doesn't seem too great a stretch to argue that the kind of visualization in which one looks and discovers is a kind of communication, even if it is only with oneself!

It is certainly true that most acts of visualization share all the characteristics of other communications activities, including having senders, receivers, messages, media and the possibilities of success or failure. Many of the acts of visualization addressed by Bishop and Lange are of the kind often employed by landscape planners and designers, in the process of trying to communicate, or understand, or explore, what the Harvard planner Carl Steinitz calls 'alternative futures'. This role of visualization – helping oneself and others to envision and understand an intended or imagined reality, often different from the present, through one or more images, still or moving – makes special demands on the senders and the technologies of the communication. The agent (person, or computer program, etc.) who prepares the visualization must first grasp the questions at hand, and the context in which they are being asked; and then must make a series of crucial decisions about such things as abstraction levels, symbolism, realism, scale and format, among others. These are the required decisions about any visual representation, and they are made all the more urgent because the representation is not purely artistic, like a painting or a poem, but is rather embedded in a real-world context, often with social, ethical, economic, political and other implications. These real-world demands are part of what make the art and science of landscape visualization so important. And that's why the contributions of this book are so timely and important as well.

In gathering in this volume concepts, technologies, applications and details from their own and others' experience, Bishop and Lange have provided a previously unavailable compendium of visualization theory and practice. Many of the techniques of computer graphics used or described have been well established heretofore and are under active development, but the particular demands of their application to environmental visualization are still being explored. In our book *Landscape Modeling: Digital Techniques for Landscape Visualization* (McGraw-Hill 2001), Hope Hasbrouck and I laid out a conceptual framework for approaching these unique demands, and described in isolation a number of the specialized modelling and visualization techniques required for constructing virtual landscapes. But we did not, as the present volume does, present a range of more integrated applications from this emergent and important field of activity. This book provides some essential updating of information in our own previous book, and broadening with application-based perspectives, as well as bringing its authors' own insights and expertise.

In the case studies in Chapters 6 to 9, the contributing authors describe applications in forestry, agriculture, energy and the urban milieu, in which computer-aided visualization is used to answer questions ranging from 'How will [it] function?' to 'What might [it] look like?' Implicit in these discussions is the critical role of determining exactly what the questions are; as in any research and communications enterprise, much can hinge on minor differences of emphasis or interpretation. The broad range revealed in the kinds of questions asked, and answered, as well as the technologies brought to bear in the process, is one of the greatest values of this work. Readers, whether professionals engaged in making such representations, or citizens increasingly involved in evaluating and responding to them, have much to learn from the selection of examples and experiences provided.

As several of the case studies demonstrate, visualization can be a powerful problem-solving tool. Because the human eye–brain system is so sophisticated in pattern recognition, difference detection, and so on, visualization can be an effective aid in detecting correlations, implications and anomalies – not just in rendering aesthetic verdicts. The scientific community has come to depend upon 'scientific visualization' in a host of non-spatial domains (mathematics, biochemistry and others) to make visible otherwise invisible phenomena. The great power of simply switching representations is a well-known problem-solving tool – one reason why, as the great theoretician Herbert Simon wrote, 'a picture is sometimes worth a thousand words'.

Aesthetics, of course, are important, too. Most of us would choose beauty in our environment – or at least not abject ugliness – whenever we could, and talented landscape architects and others can make the best of many engineering, design and planning problems by providing solutions both functional and aesthetic, whose effectiveness can be partly tested through visualization. Although shared aesthetic judgments are impossible to arrive at, and are certainly harder than shared objective evaluations of, say, the visibility of objects at a distance, or the effectiveness of camouflage in natural environments (both questions to which visualization tools can be effectively employed for answers), the great value of the techniques described herein is their ability to augment the process of public and personal decisions with essential visual information. This book provides ample evidence of the power of the art and science of computer-aided visualization to assist in decision making across a range of critically important human activities, literally shaping the face of the earth for generations to come.

Stephen M. Ervin
Harvard Design School

CONTRIBUTORS

Steve Barnhart is a founding principal landscape architect with SKB Associates located in Guelph Ontario. He holds a BSc in geology from the University of Waterloo and a MLA from the University of Guelph. His professional practice includes many projects using advanced visualization technologies. He has also conducted a number of applied research projects utilizing web-based tools and methods.

John Benson (1947–2004) was Professor of Landscape Planning and Management and Director of the Landscape Research Group at the University of Newcastle. He was also Editor-in-Chief of the *Journal of Environmental Planning and Management*.

Duncan Cavens is a researcher at ETH Zürich (the Swiss Federal Institute of Technology), where he works on the simulation of landscape change using agent-based techniques. He graduated from the University of British Columbia with a degree in landscape architecture and an interdisciplinary Master's in forestry, landscape architecture and computer science.

Liviu Coconu graduated in 2001 from the 'Politehnica' University of Bucharest, and he is now working on his PhD at the Zuse Institute Berlin (ZIB).

Alastor Coleby is an environmental geoscience and planning graduate with experience of environmental assessment and management systems. He is studying for a PhD at Heriot Watt University on public perceptions on the design of wind turbines and methods of public participation in the development of windfarms.

Carsten Colditz graduated in 2003 from the Dresden University of Technology, and he is now working on his PhD at the University of Constance.

John Danahy is Director of the Centre for Landscape Research (CLR) and a studio professor in landscape architecture at the University of Toronto. He has led the development and application of real-time visualization, immersive and collaborative software for urban landscape design at CLR since the early 1980s.

Anne-Marie Davies is an environmental economist. She worked in the Scott Sutherland School from 1999 until 2003 as the principal researcher on both the 'Streetscapes' and EC 'Greenspace' projects. Anne-Marie is currently working as a consultant with the New Economics Foundation in London.

Oliver Deussen received his PhD in 1996 at the University of Karlsruhe. From 1996 until 2000 he was research assistant at the Otto-von-Guericke University of Magdeburg. From September 2000 until March 2003 he was professor for computer graphics and media design at Dresden University of Technology, and since April 2003 he has been full professor at the University of Constance. He has written more than 60 reviewed articles and is author or co-author of several books in the field of computer graphics.

Ben Discoe is a veteran of the Virtual Reality industry. He worked at Sense8 on the 3D/VR API WorldToolKit, and at Intel's Research Labs, where he prototyped three-dimensional applications to catalyse the formation of future software industries. Ben currently runs the Virtual Terrain Project, an Open Source suite of software for rapid construction of interactive three-dimensional scenes from any geospatial data, with a web site acknowledged as the leading information source on the subject.

John C. Ellsworth, FASLA, CLARB holds a Master of Landscape Architecture degree and is a Professor in the Department of Landscape Architecture and Environmental Planning at Utah State University. He is also President of Ellsworth and Associates, landscape architects, inc. (www.ellsworthandassociates.com).

Stephen M. Ervin is Director of Computer Resources and Assistant Dean for Information Technology at the Harvard University Design School, and is a Lecturer in the Department of Landscape Architecture. Ervin teaches and conducts research in the areas of design, computing, media and technology, with a special interest in landscape modelling and visualization. He is the founding chairman of ASLA's Open Committee on Computers in Landscape Architecture.

Rebecca Ford is a PhD student in the School of Resource Management at the University of Melbourne. She has a background in forestry and most recently worked with the Australian Greenhouse Office.

Issa A. Hamud, PE is a professional engineer with Bachelor and Master's degrees in engineering from Utah State University. He is director of the Department of Environmental Health, city of Logan, Utah.

Hans-Christian Hege is director of the Scientific Visualization department at Zuse Institute Berlin (ZIB). He finished his theoretical physics studies at Free University Berlin in 1984. He co-founded mental images (1986) and Indeed – Visual Concepts (1999) and joined ZIB in 1989. He has published more than 140 reviewed articles, is co-editor of three monographs and co-editor of the book series 'Mathematics+Visualization' at Springer.

Sigrid Hehl-Lange is senior researcher in the landscape research group at the Institute for Spatial and Landscape Planning at the Swiss Federal Institute of Technology (ETH Zürich). She graduated at the TU Berlin with a Dipl.-Ing. degree in landscape planning. She received her PhD from the Department of Civil, Environmental and Geomatic Engineering at ETH Zürich. Her main fields of research are ecological planning, landscape ecology, GIS and landscape visualization.

Adrian Herwig studied landscape planning at Essen University (Germany). Since the mid-1980s, he has been using computers for the visualization of planning. He now runs his own landscape architecture firm in Stechau, Germany where he uses game-based visualization tools for garden design, landscape planning, village development planning and much more. Since 2000 he has been working with the *Lenné3D* project.

Einar Kretzler is professor for informatics in landscape architecture at the Anhalt University of Applied Sciences (Germany) and runs his own visualization firm. He studied landscape architecture at the University Hannover.

Michael Kwartler is an architect, planner, urban designer and educator. He is the founding director of the Environmental Simulation Center, a non-profit research laboratory created to develop innovative applications of information technology for community planning, design and decision making. He conceived and directed the design and development of CommunityViz, the first GIS-based planning decision support software to fully integrate virtual reality with scenario design, impact analysis and policy simulation.

Richard Laing is a quantity surveyor and senior lecturer at The Scott Sutherland School at Robert Gordon University, Aberdeen. He was the project manager for Streetscapes, as reported in this book. The methods used in that project are now being further developed through the EC-funded 'Greenspace' project.

Daniel Loiterton is a PhD student in the Department of Geomatics, at the University of Melbourne. He is using virtual environments and agent-based models to explore visitor movements in urban parks and gardens.

Andrew Lovett is a senior lecturer in the School of Environmental Sciences at the University of East Anglia (UEA), Norwich, UK. His research interests include the use of GIS and visualization techniques in landscape planning and he is currently developing a new virtual reality facility at UEA.

Abraham N. Medina, ASLA holds a Master of Landscape Architecture degree from Utah State University. He is employed at Eggers Associates, Landscape Architecture in Ketchum, Idaho.

Michael Meitner, an environmental psychologist by training at the University of Arizona, is currently an assistant professor in the Department of Forest Resources Management at the University of British Columbia. His research interests include visual perception, scenic beauty and aesthetics, data/environmental visualization, Geographic Information Systems and human emotional processing.

David Miller is Acting Head of the Landscape Change Science Area of the Macaulay Institute. He also holds an honorary professorship in the OPENSpace Research Centre at Heriot Watt University. His areas of research include quantitative and qualitative assessments of the effects of drivers of change on the visual landscape.

Jane Morrice holds a degree in economics, and is a research scientist in the Landscape Change Science Area of the Macaulay Institute, working on the geographical modelling of landscapes and public participation in landscape planning within the *VisuLands* project.

Eihachiro Nakamae is the chairman of Sanei Co. Ltd. His research interests include computer graphics and road design CAD. He was granted the title of emeritus professor from both Hiroshima University and Hiroshima Institute of Technology, and received ME and Dr Eng. from Waseda University.

Brian Orland is Head of Landscape Architecture at Penn State University. He has taught landscape design and land resource evaluation. Research interests include the computer modelling of environmental impacts and the design of online information systems to support community-based planning initiatives. Studies have included the impacts of highway development, of military training, and of insect pest, harvesting and fire impacts on national forests.

Philip Paar is scientific assistant at the Zuse Institute Berlin (ZIB) and coordinator of the *Lenné3D* project at the Leibniz-Centre for Agricultural Landscape and Land Use Research (ZALF). He studied landscape planning at the Technical University Berlin.

Nathan H. Perkins is an associate professor in the School of Environmental Design and Rural Development at the University of Guelph, Canada. His research is in the area of understanding person–environment interactions, specifically those of children and hospital patients. He has been using digital technologies for over a decade to facilitate the collection and analysis of information from these special populations.

Wayne Piekarski is the assistant director of the Wearable Computer Laboratory in the School of Computer and Information Science at the University of South Australia. Dr Piekarski's qualifications include: BE in Computer Systems Engineering and PhD in Computer Science from University of South Australia.

Xueying Qin is an associate professor in Zhejiang University, People's Republic of China. Her research interests include computer graphics, image processing and computer vision. She received her PhD from Hiroshima University, Japan, and MS and BS from Zhejiang University and Peking University, People's Republic of China, respectively.

Jörg Rekittke is a lecturer in landscape architecture at the University Siegen and teaching assistant at the Faculty of Architecture, RWTH Aachen University, Department of Urban Design and Regional Planning. He holds a doctoral degree from the RWTH Aachen University. He studied landscape architecture at the Technical University, Berlin and at the ENSP, Versailles.

Jon Salter is a graduate student with the Collaborative for Advanced Landscape Planning at the University of British Columbia, where he is working on the evaluation of landscape visualization and information visualization for public involvement in land-use planning. Jon's background includes degrees in both psychology and natural resource conservation.

Stephen Scott is an architectural technologist and research assistant at The Scott Sutherland School. He specializes in computer visualization of the built environment. Stephen is currently working on research which aims to develop the manner in which interactive virtual environments can be used within the social sciences.

Stephan Sheppard is associate professor in Forest Resources Management and Landscape Architecture at the University of British Columbia, and conducts research in collaborative planning, landscape perception and visualization. He directs the Collaborative for Advanced Landscape Planning (CALP), an interdisciplinary research group using immersive/interactive visualization techniques to support public communications, sustainable land use planning and resource decision making.

Christian Stock studied physics at Hamburg University before moving to Wellington, New Zealand. There he studied at Victoria University for his PhD in geophysics, which he received in 2001. Currently, he holds a position as a research fellow in the Department of Geomatics at Melbourne University.

Katsumi Tadamura is a professor in the Faculty of Engineering at Yamaguchi University. His research interests include photo-realistic rendering, physically based animation, scientific visualization and graphical user interface. He received an ME and a PhD in system engineering from Hiroshima University.

Bruce Thomas is the director of the Wearable Computer Laboratory in the School of Computer and Information Science at the University of South Australia. Research interests include: wearable computers, user interfaces, augmented reality and virtual reality. Dr Thomas' qualifications in computer science include an MS from University of Virginia and PhD from Flinders University.

Liisa Tyrväinen works as a project director at the University of Helsinki, Department of Forest Ecology. One of her main research interests has been to develop methodologies for integrating scenic values to forest planning. The approaches employed include landscape preference studies, social surveys, scenic preference models and visualization.

Jori Uusitalo is an acting professor in forest engineering at the University of Joensuu, Finland. He has studied forest visualization at the Imaging Systems Laboratory, University of Illinois, and worked earlier as an acting professor and senior lecturer in Wood Technology at the University of Joensuu. He currently coordinates several research projects that deal with wood procurement logistics, harvester operators' working techniques and modern tree bucking systems.

Kathryn Williams is a lecturer in the School of Resource Management, University of Melbourne. Her research and teaching is concerned with the psychology of natural resource management, with particular emphasis on landscape cognition and experience.

ACKNOWLEDGEMENTS

First and foremost we must thank the many authors who have contributed to this volume. We approached the people that we believe to be at the cutting edge of visualization development and environmental applications. All gave their time generously in preparing their sections of the text and were invariably willing to bend and adapt their contributions at the whims of the editors. In particular, we acknowledge the contribution of John Benson who died between the time he submitted his manuscript and completion of the book. Neither of us had the chance to meet John personally, but we greatly appreciate his contribution not only to this volume but also to landscape research and landscape education.

Where one of our contributing authors has included his or her images in their section of the book we have not acknowledged each individual image. Unless otherwise identified below, the images belong to the author(s) of the section of text in which the figures are referenced. However, we have also drawn a number of images from other places to illustrate particular aspects of visualization and its development. In these cases we would like to thank the following for permission to include their work.

Figure 1.2(a) Jim Siebold (http://www.henry-davis.com/MAPS/AncientWebPages/121.html)

Figure 1.2(b) Bremner & Orr Design Consultants Ltd, Tetbury, England © 2004

Figure 1.6(a) and (b) Michael Kwartler, Environmental Simulation Center, New York

Figure 1.9(a) and (b) DHM25 © 2004 swisstopo (BA045986)

Figure 1.9(c) reprinted from *Landscape Planning*, 4, Myklestad, E. and Wagar, J.A. PREVIEW: computer assistance for visual management of forested landscapes, 313–331, copyright (1977) with permission from Elsevier

Figure 1.9(d) Scott Arvin, Donald House, Midori Kitagawa and Greg Schmidt, Texas A&M Visualization Laboratory, copyright 1998

Figure 1.10(b) US Army Photo by Ron Carty

Figure 1.11(b) Wayne Piekarski and Bruce Thomas, University of South Australia

Figure 2.1(a) US Geological Survey

Figure 2.1(b) Hellmuth Obata + Kassabaum Inc. (project design and three-dimensional modelling) and Advanced Media Design Inc. (VIZ rendering)

Figure 2.2 Alan H. Huber, US EPA

Figure 2.3 Frank Hardisty, University of South Carolina

Figure 2.4 Natalia and Gennady Andrienko, Fraunhofer Institute AIS, CommonGIS, is available for commercial and free non-commercial use at www.commongis.de

Figure 2.7 Reprinted from *Landscape and Urban Planning*, 54, Krause, C.L. Our visual landscape: managing the landscape under special consideration of visual aspects, 239–254, copyright (2001) with permission from Elsevier

Figure 2.8 Reprinted from *Landscape and Urban Planning*, 54, Hehl-Lange, S. Structural elements in the visual landscape and their ecological functions, 105–113, copyright (2001) with permission from Elsevier

Figure 2.9(a) Aerial Photography & Digital Elevation Model: © Amt für Gewässerschutz und Wasserbau des Kantons Zürich and Institut für Umwelt-Risiko-Management (ITR), Image Processing and Computer Graphics: Remote Sensing Laboratories (RSL), Department of Geography, University of Zürich

Figure 2.9(b) Reprinted from *Landscape and Urban Planning*, 54, Perrin, L., Beauvais, N. and Puppo, M. Procedural landscape modeling with geographic information: the IMAGIS approach, 33–48, copyright (2001) with permission from Elsevier

Figure 2.10 Edward Verbree, OTB Research Institute for Housing, Urban and Mobility Studies, Delft

Figure 2.11 William Winn, University of Washington, Seattle

Figure 3.1 Image from the DMAP website (www.dmap.co.uk), courtesy of Alan Morton

Figure 3.6 Cliff Ogleby, University of Melbourne

Figure 4.9 DHM25 © 2004 swisstopo (BA045986)

Figure 4.11(a) Photo courtesy of SEOS Ltd

Acknowledgements

Figure 4.12(a) Cliff Ogleby, University of Melbourne. Screen shot reprinted with permission from Microsoft Corporation

Figure 4.12(b) Screen shot reprinted with permission from Microsoft Corporation

Figure 4.13 Eric Champion, University of Melbourne. Adobe product screen shot reprinted with permission from Adobe Systems Incorporated

Figure 4.14 Philip Paar, Zuse Institute Berlin and Erich Buhmann, Anhalt University of Applied Sciences

Figure 5.1 Ken Fairhurst, Collaborative for Advanced Landscape Planning (CALP), University of British Columbia

Figure 5.2(a) Keoysan Seyfarth and Associates and the California Department of Water Resources

Figure 5.5 Diversified Animated Technologies Associates, Inc. (DATA), in association with Wickerworks Video Productions, Englewood, CO, courtesy of US Central Federal Lands Highway Division

Figure 7.3 Screen shot reprinted with permission from Microsoft Corporation

Figure 7.8 Cartoons by Mele Brink and Jörg Rekittke

Figure 8.1 David Watson, freelance IT/CAD consultant to the landscape profession

Figures 8.2, 8.3 and 8.4 are copyright © 2004 Ellsworth and Associates, landscape architects, inc., all rights reserved

Figures 8.5, 8.6 and 8.7 are copyright © 2003 Abraham Medina, all rights reserved

Figures 8.9(a) and (b) The road networks are reproduced with permission of the Ordnance Survey, under licence to The Macaulay Institute No. GD272825G 2001

Figure 9.20(a) Screen shot reprinted with permission from Microsoft Corporation

Figure 10.12 Screen shot reprinted with permission from Microsoft Corporation

Figure 10.15(b) ArcView Graphical User Interface is the intellectual property of ESRI and is used herein with permission. Copyright © ESRI

Thanks also to Deb Thomas who contributed care and dedication to several editing stages.

Part I

Understanding visualization

CHAPTER 1

COMMUNICATION, PERCEPTION AND VISUALIZATION

Eckart Lange and Ian D. Bishop

Why visualize?

Humans perceive their environment through their senses. Commonly these are distinguished as an auditive system (the sense of hearing), a tactile system (the sense of touch), a kinaesthetic system (the ability to sense and coordinate movement), a vestibulary system (the sense of balance), an olfactory system (the sense of smell), a gustatory system (the sense of taste) and a visual system (the visual sense). Vision is easily the dominant component. This is underlined in everyday language by phrases such as seeing is believing and a picture is worth a thousand words, also a grand idea is a vision and we envision the future. Indeed, Bruce *et al.* (1996) estimated that 80 per cent of our impression of our surroundings comes from sight.

Because of our long history of drawing pictures of our environment and the important entities within it, we hardly need to answer the question 'Why visualize?' On the other hand, technology is giving us wholly new ways to visualize and to use visualization. This makes it worth reiterating some key points.

- We want visualization to give us the opportunity to see, experience and understand environmental changes before they occur.
- Through the ability to share this experience and potential for exploration, visualization will help communities (of whatever size) to build consensus and make decisions about their future.
- The relationship of people to their environment is a key contributor to environmental decisions and visualization can help us learn more about that relationship.

Being able to visually represent the existing real world as well as potential alterations is essential for landscape designers and planners to express and communicate their thoughts. In the past, maps, plans and sections have been predominately used. These representations are at a high level of abstraction. For the understanding of both the general public and the experts, it is important to communicate a proposal in perspective view, providing a more natural and direct approach to communication.

A brief history

For several thousands of years people have been using images to convey information. Traditional analog visualization techniques for the representation of concepts in planning and design are plans, sections, sketches, perspective drawings, photomontages, and physical models. Of these techniques, physical models, such as the ones found in Egypt and early Chinese tombs, and sketches are the oldest (Zube *et al.* 1987). Examples of drawings date back beyond 30 000 BC as illustrated by the charcoal drawings in the cave of Chauvet-Pont-d'Arc (Ministère de la Culture et de la Communication 2003).

Despite the fact that these visualizations were produced thousands of years ago they are sometimes so well preserved that we can even extract information about the state of the environment at the time. Plants and animals depicted in prehistoric rock art in the Sahara region provide impressions of a long gone, much more humid landscape quite different to our mental image of the Sahara today (Figure 1.1)

Many of these early cave paintings represent the important animals of the local environment at the time. For whatever reason, the ancient visualizers chose to concentrate on specific components of their environment (e.g. horses and bison) while only very occasionally providing representation of other animal groups such as birds and fish. This selectivity has been a feature of the visualization process ever since.

Mapping is known to have commenced at least 8000 years ago. One of the earliest known maps is from Turkey, dated at around 6200 BC

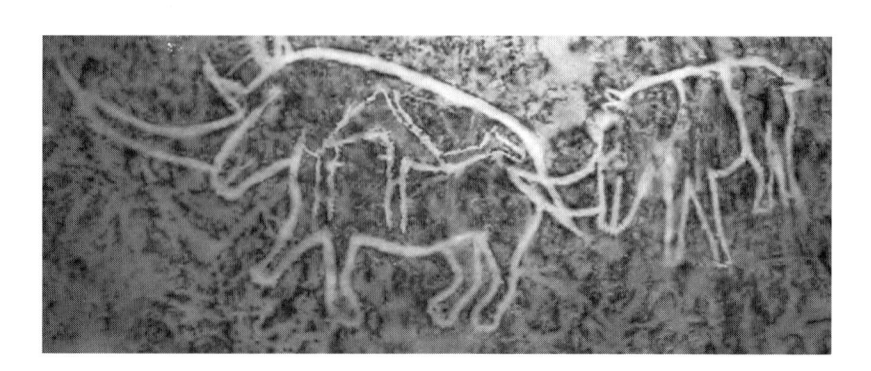

1.1 Rubbings of prehistoric carvings from the Sahara showing both rhinoceros and camel.

Communication, perception and visualization

1.2 The use of a pseudo three-dimensional approach has been common in mapping for hundreds of years: (below) Palestine 565 AD; (right) St Andrews, Scotland 2004 AD

(Delano-Smith 1987). Because of its similarity to the layout of the excavated houses, this wall painting is interpreted as a portrayal, in plan, of the old settlement of Catal Hüyük.

The first known examples of terrain representation date from more than 4000 years ago, scratched onto clay tablets (see Delano-Smith 1987). The mapping of Ga-Sur (now Yorghan Tepe), near Kirkuk in present day Iraq (dated 2300 BC), illustrates the earliest known examples of a topographic map with a correct orientation. A little later, a scale drawing of the town of Nippur (ca. 1500 BC) shows that the Babylonian cartographers practised the essential principles of mapmaking as we know them today (Delano-Smith 1987). Between that time and the second half of the twentieth century these two-dimensional representational techniques were refined and our knowledge of the world expanded rapidly but the essential form of maps remained the same. However, cartographers were also seeking a more natural interpretation of spatial arrangements through a pseudo three-dimensional representation (Figure 1.2).

Developments in three-dimensional representation came during this period as much from the world of art as from science. The technology of perspective had a substantial impact in the early fifteenth century. Although invented around 465 BC in Greece (as proved by skenographic representations, in Geyer 1994), it was several hundred years before the perspective was reinvented in the Renaissance, and became a

Understanding visualization

1.3 An analogue technique for deriving perspective (from Dürer 1525)

common tool for the presentation of final architectural designs. The oldest known perspective from this period is a mural dated after 1317 by Giotto in the Bardi Chapel of Santa Croce in Florence. The Italian artist and architect Filippo Brunelleschi (1377–1446) further developed the technique of perspective drawing. He established the use of a 'vanishing point', an imaginary single point on the page at which all the parallel lines meet. The use of the vanishing point was particularly important as three-dimensional representation of landscapes developed. In the sixteenth century, Albrecht Dürer wrote an introductory manual of geometric theory for students (Figure 1.3). This includes the first scientific treatment of perspective by a Northern European artist (Dürer 1525).

Within the discipline of landscape architecture Humphry Repton (1803) pioneered visualization applied to design issues on a site and on a landscape scale. In his famous Red Books, which he used to show his designs to his clients, he concentrated on the representation of proposed changes in the landscape in perspective view. The existing situation could be compared with the proposal by having the proposal drawn under a movable flap (Figure 1.4). Purposefully, he made little use of two-dimensional representations. A variation of this before-and-after technique was used by Frederick Law Olmsted for communicating his famous design for Central Park, New York (Beveridge and Schuyler 1983).

A specific branch of drawing and painting is the panorama. It was invented and patented in 1787 by Robert Barker (1739–1806) (Oettermann 1980; Comment 2000). In another crossover between art

1.4 Repton's approach to before and after landscape representation (images reproduced in the colour plate section)

and science in the generation of landscape simulations, what we today consider an art form was then judged as a technical and scientific idea. The first panorama drawn by Barker illustrated the view from Carlton Hill overlooking Edinburgh. Independently, the artists, engineers and scientists Hans Conrad Escher von der Linth (1767–1823) and Johann Adam Breysig (1766–1831) also produced their first panoramas. Documentary panoramas in small formats were frequently produced to realistically portray a contemporary or historical urban or rural landscape. In the nineteenth century the monumental panorama became an early mass medium attracting numerous visitors: the equivalent to today's Imax theatres.

To show the panoramas, specifically constructed buildings (rotundas) were set up in many major cities of central Europe and North America. Popular motifs were martial scenes and war victories but, equally, panoramas of well-known cities, spectacular landscapes and exotic travel destinations were on display.

With the rise of photography, which was easier to use and much faster to produce, the panorama lost popularity in the middle of the nineteenth century. Nowadays, most panoramas have been destroyed. The oldest remaining example, painted by Marquard Wocher between 1809 and 1814 on a canvas of 7.5 m × 39 m, is in Thun, Switzerland. The panorama of the battle of Murten (1880) was recently reconstructed (at great expense) for the Swiss Expo of 2002. In a giant rusty cube designed by Jean Nouvel amidst the Lake of Murten and only accessible by boat, in a panoramic multimedia installation, the real panoramic view across the lake and the historic Murten panorama could be compared (Figure 1.5). This mixing of media suggests a renewed interest in panorama as a form of communication and entertainment. Photographic techniques and especially the emergence of digital photography have allowed fast and easy production of photographic panoramas. The benefit of the digital panorama is that it can be seen in a wide variety of sizes, from a scrolling window on a computer screen to projection-based panoramas just as large as their painted forebears.

Understanding visualization

1.5 The battle of Murten (above) as a painted panorama and (left) its special exhibit space in the lake

The scientific approach to perspective made it easy, several centuries after Giotto and Dürer, for computers to create perspective drawings. Early three-dimensional representations in computer-aided design (CAD) were restricted to parallel projection (commonly isometric). However, some very early approaches to landscape representation were drawn in perspective. Myklestad and Wagar (1977) used minimalistic tree symbols in order to show proposed landscape changes caused by timber harvesting and regrowth through the silvicultural rotation (see also Figure 1.9). Also at the end of the 1970s, the Defense Mapping Agency in the USA began to develop visualizations to support advanced aerospace systems. Examples with shaded relief, draped imagery from remote sensing and shaded building volumes (Faintich 1980) marked the beginning of a dramatic period of development in the style of landscape representation. The emphasis during this period was on static display options. A notable exception was the work of Molnar (1986) who argued the case for real-time exploration of landscape designs. He was, however, limited to wire-frame representation.

Animation of simulated landscapes for planning purposes did not begin with computer graphics (Zube *et al.* 1987). Working in the analog rather than digital world, the pioneers of dynamic simulation came from the Berkeley Environmental Simulation Laboratory (Appleyard and

1.6 The three-dimensional model as a visualization environment: (right) a view of an urban streetscape (with manual editing); (far right) the gantry system which carried the camera along defined paths

Communication, perception and visualization

1.7 The physical model shows well the terrain surface but cannot portray the typical atmospheric conditions of the site: (above) real landscape, Großer Windgällen; (left) physical model by Eduard Imhof

Craik 1978). Among analog techniques, only the physical model permits free, i.e. dynamic, eye movement of the observer. Using a miniature camera hung from overhead gantries (Bosselmann 1983), models can be explored at eye level. The camera system was called an endoscope or modelscope and gave the user freedom of movement through their urban model (Figure 1.6). However, even a very precise physical model with detailed objects and exact contouring cannot capture the visual appearance of an environment completely. The presence of haze, for example, cannot be simulated in a physical model, as seen in Figure 1.7.

For some time, the photomontage has been a common technique for creation of images of changing landscapes (e.g. Bureau of Land Management 1980). In the 1980s, computer graphics cards and software emerged which allowed photographs to be captured, stored and manipulated. Elements could be cut from one image and pasted into another. As a result, the analog photomontage was quickly superseded by the digital photomontage (Lange 1990; Orland 1988). Unlike the traditional analog photomontage, a digital photomontage can reach a relatively high level of geometric accuracy. This can be achieved by superimposing three-dimensional vector data over the two-dimensional image data (Figure 1.8).

Meanwhile, three-dimensional presentation of spatial data using computers was developing at the Laboratory for Computer Graphics and Spatial Analysis at the Harvard Graduate School of Design. Dougenik (1979) and Hanson and Lynch (1979) showed census and other demographic data as a pseudo terrain surface. Developments in this period also laid the foundation for further development of recently established GIS and image-based software products (Dangermond *et al.* 1981) establishing the need for three-dimensional visualization capabilities in standard commercial products (Faust *et al.* 1981).

Understanding visualization

1.8 The expansion of an existing reservoir (left) is simulated by placing a CAD perspective over a photograph (right)

This was the basis for the later explosion of interest in scientific visualization that emerged in the 1980s with the realization that computer graphics – and three-dimensional representation in particular – could display complex data in ways previously impossible (McCormick *et al.* 1987). The new display options permitted scientists to find relationships or understand structures that had been obscure. Initially the development of new visualization techniques and applications was based on large and expensive super computers and specialized graphics workstations (Ribarsky *et al.* 1994). However, almost immediately similar capabilities began to appear on typical desktop computers (Peltz and Kleinman 1990).

Applications quickly followed in areas such as polymer chemistry (Zoll and Rosenberg 1990), ecology (Hamilton and Flaxman 1992) and planetary science (Moore 1990). The environmental management discipline also realized in the early 1990s that there was great potential in the use of visualization technologies. Areas of specific application included pollution management (Rhyne *et al.* 1993), earth sciences (DiBiase 1990), bushfires (French *et al.* 1990) and natural resource management (White 1992). At the same time, cartographers started early in exploiting the new technology, particularly in the animation of historical data (MacEachren and Ganter 1990; DiBiase *et al.* 1992).

Three-dimensional modelling for landscape visualization continued to develop in sophistication throughout this period (Figure 1.9). As the technology developed researchers became increasingly concerned with the validity of visual simulation techniques for representing environmental change and assessing public reactions. The first efforts to establish image validity were in image manipulation (Bishop and Leahy 1989; Vining and Orland 1989). In three-dimensional modelling, Oh (1994) studied perceptual responses to campus simulations. In the natural environment, Bergen *et al.* (1995) tested the validity of computer-generated

Communication, perception and visualization

1.9 The change from the representational styles developed in the 1970s to the far more elaborate and realistic landscapes of the 1990s: (top) landform using distorted squares; (top right) landform with orthophoto overlay; (above) the original tree representations; (right) highly detailed tree models (images reproduced in the colour plate section)

forest environments. More recently, Lange (2001a) systematically tested the role of detail. Bishop and Rohrmann (2003) compared responses to a real walk in an urban landscape with responses to animations. The issue of simulation validity is considered further in Chapter 5.

The idea of linking visualization to geographic information systems also emerged at the end of the 1980s (Lang 1989; Buttenfield and Ganter 1990). The potential to develop three-dimensional models, and hence landscape renderings, from GIS data in an automated process is rapidly expanding access to landscape simulation (Hoinkes and Lange 1995; Perrin *et al.* 2001; Ribarsky *et al.* 2002). Appleton *et al.* (2002) have reviewed the software and the issues involved in making these connections. In Chapter 3 the opportunities for building three-dimensional models from GIS and other digital data sources are reviewed in detail.

The most recent development to shift our thinking on the potential of visualization is the sudden ubiquity of the Internet. With transmission speeds increasing (broadband) it has become convenient to share three-dimensional models publicly. One established technology to enable the presentation of three-dimensional models on the Internet is VRML (Virtual Reality Modelling Language). The use of VRML-models in landscape and environmental planning provides easy access to planning concepts expressed in three-dimensional form (Nadeau 1999; Lange 2001b; Lovett *et al.* 2002). Using a three-dimensional VRML-model, interested citizens are offered much more than the predetermined viewpoints often used in reports. They can view proposals from any viewpoint they like. Hotlinks to data can be associated with objects and a certain degree of interactivity with model components can be implemented. Other three-dimensional web standards are emerging which offer further functionality (see also Chapter 4, p. 74). The extension of automated, interactive, model-based systems to the web environment can provide a base for a quantum leap in public participation in planning (see examples in the applications chapters of Part Two and also Chapter 10 – Visualization prospects).

Towards alternative 'realities'

We think of virtual reality technology (VR) as a very recent invention. However, in 1962, a prototype was developed by Morton Heilig, a private inventor. His system, which he called Sensorama, simulated a motorcycle ride through New York. In addition to a visual representation, he incorporated other sensual experiences including fan-generated wind and the smells and noise of New York. Because the route was pre-recorded there was no possibility for the user to interact with the system.

Only a few years later, Ivan Sutherland (who pioneered several areas of computer graphics including flight simulation) developed the first head-mounted display (Sutherland 1968). This output device, which together with the data glove by Jaron Lanier (Zimmerman and Lanier 1987), was for many years considered essential for an ultimate experience of immersive virtual reality. Several definitions of virtual reality exist (see also Table 2.1) but it is broadly characterized as a computer-generated, three-dimensional environment providing interactivity and immersion (Gaggioli 2001). Because of the absence of interactivity Sensorama would not be classified as virtual reality today.

The shift from the film/videos and models, first used in flight simulators after the Second World War, towards digital landscape representation provided great impetus to the advancement of virtual reality technology. Nowadays, VR is used not only in aviation, but also in industry and research within diverse fields including medicine, chemistry, mechanical engineering and architecture. It is finding increasing application in landscape architecture and landscape and environmental planning (van Veen *et al.* 1998; Verbree *et al.* 1999; Bishop *et al.* 2001; Fairbairn and Taylor 2002).

Danahy (2001) stressed the importance of an immersive environment arguing that the 'dynamic qualities of looking around, ... using one's peripheral vision, and focusing with foveal vision on objects of attention are fundamental to a person's visual experience in landscape' (p. 125). Panoramic projections are one option for filling the field of view and generating immersion in the scene. This reminds us of the popularity of painted panoramas in the nineteenth century. Digital three-dimensional data can be used to compute panoramas of historic and present landscapes (Rickenbacher 2001a) as well as proposed changes (Lange *et al.* 2001). Using high-resolution data, state-of-the-art printing or projection technology, permits production of monumental panoramas even larger than the original rotundas (Rickenbacher 2001b). These models can also be viewed using colour filters, polarization glasses or LCD glasses to make the experience stereoscopic. In general, with the increasing availability of digital data and continuously falling costs for projection technology and large-format printing we anticipate a continuing revival of the panorama. A current, very popular example is the 106 m × 32 m panorama of Mount Everest realized in a mixed technique of digital photography and digital paintbrush. It is on display in the Leipzig gasometer, Germany (Asisi 2004).

A complete surrogate landscape experience should also ideally involve natural movement including expenditure of energy and haptic feedback from objects encountered. The importance of interactivity as a part of the virtual landscape experience was stressed by Bishop and Dave (2001). Driving simulators are widely available for training purposes but seldom used for simply moving through the landscape (Figure 1.10). Other researchers have used a bicycle or tricycle as the mode of virtual transport (Allison *et al.* 2002), while walking can be accommodated by the omni-directional treadmill (Darken *et al.* 1997) or roller-blades (Iwata and Fujii 1996). Also available are force feedback devices simulating the haptic experience of opening a virtual door, pushing a virtual button or using a virtual scalpel in a distributed medical VR environment.

Further extensions of the virtual reality concept will include development of remote collaboration and decision making within the virtual environment. By linking two (or more) immersive facilities via a high-speed connection, support can be provided for fully three-dimensionally rendered virtual humans, motion and speech in real time within an immersive environment (blue-c.ethz.ch). This approach is already widely used in online computer games where multiple players in different parts of the world share and interact in common game-space.

Instead of trying to completely simulate the sensual experience, which will always be difficult because of the complexity of our environment,

1.10 Examples of navigation options in virtual environments: (top) a simulated tram ride – here using video but could also be computer-generated imagery; (above) movement through the landscape using an omni-directional treadmill and panoramic projection system

Understanding visualization

1.11 New directions in alternative realities: (right) Blue-C environment for remote collaboration; (below) real-time augmented reality

augmented reality (AR) follows a different approach (Hedley *et al.* 2002; Thomas and Piekarski 2003). AR combines real and virtual objects in a real environment by correct alignment, occlusion and lighting of virtual objects in the real view. According to some authors (Azuma *et al.* 2001) it runs in real time while other researchers (Nakamae *et al.* 2001) have worked with video recordings of the real world. In contrast to VR applications, the user does not experience a merely synthetic world. This way the user can experience the real world with changes shown by computer-generated objects ('augmentations') superimposed or interpolated (Figure 1.11).

Another interesting approach is pursued by Pair *et al.* (2003) in their FlatWorld project. Inspired by Hollywood set-design techniques, a multisensory mixed reality (MR) experience is created by combining virtual and real elements such as doors and windows which can be touched and physically opened, then providing a view out to a virtual environment or another room which can be entered.

AR/MR systems are currently developed mainly for the military and medical sectors. In planning they will allow us to study proposed changes to the landscape on site, permitting a complete sensual experience with a visual augmentation, e.g. by utilizing miniature displays integrated in eyewear (e.g. www.microopticalcorp.com) showing potential changes in the landscape (see also Chapter 10). A calibration followed by a dynamic superimposition procedure to register virtual objects with real objects is essential (Rolland *et al.* 2002) but not yet resolved (You *et al.* 1999). New positioning technologies (GPS and inertial navigation system), combined with object recognition, may provide a solution.

Representing landscape elements

Real landscapes are highly complex structures often covering very large areas. For visualization this is an extremely challenging task. Only within the last few years have sophisticated computer-based technological innovations allowed us to work with, and in, three and four dimensions. Looking at the real landscape, from the point-of-view of visualization, the most important variables determining the visual appearance of a landscape are terrain, vegetation, animals and humans, water, built structures as well as atmosphere and light (Ervin 2001). Depending on the issues, the planning purposes or the landscape in question, only some of these landscape elements may be present or need to be represented in high detail. However, each of these elements could be a major obstacle for achieving a representation with a high degree of realism (Lange 1999).

Terrain

Recent important developments have included very realistic and also efficient representation of terrain (Doellner and Hinrichs 2002; Nebiker 2003). As a prerequisite, digital elevation models are needed. In many countries this data is nowadays readily available. Even more important for realistic large area views is high-resolution remote sensing imagery that can be draped over a terrain model. Depending on the colour spectrum (bandwidth) this imagery can directly reflect the actual land use

Understanding visualization

1.12 Although the top image is draped with a high resolution aerial photograph there is a clear absence of important visual detail; in the image on the right each newly growing tree is modelled

information. Aerial orthophotos, which are now available at a resolution of as little as 10 cm, provide the basis for highly realistic visualizations. Furthermore, satellite-based sensors such as the recently launched QuickBird (October 2001) or Ikonos (1999) are constantly improving and are already achieving resolutions of 61 cm in the case of QuickBird (see Chapter 3 for further details on the use of digital terrain and remote sensing data). In small areas, especially when the camera is close to ground level, even very high-resolution imagery can produce unrealistic foreground effects and individual elements, such as vegetation, may need to be modelled (Figure 1.12).

Vegetation

Because of its richness in geometry, vegetation is perhaps the most challenging landscape element. Real vegetation is very complex, as it consists of a large number of objects such as leafs, buds, flowers, twigs, bark, etc. Even more challenging is the diversity of vegetation elements in the landscape. Depending on neighbourhood relationships, competition for light and nutrients or any impact caused by human intervention or natural phenomena, one particular tree species can take quite different physical forms.

Vegetation is typically represented by either applying texture maps on simple rectangular polygons, so called billboards, or by detailed polygon-based modelling of the geometry of the vegetation (see Reffye *et al.* 1988). The problem with the latter approach is that even one single tree with leaves or needles can consist of thousands or even millions of polygons. Consequently, this has a considerable impact on the time it takes the computer to draw (render) the picture ('so many polygons, so little time'). This can undermine the utility of the whole simulation process (see also Chapter 4, 'Efficient modelling and rendering of landscapes', p. 56).

Texture mapping is a very efficient and simple method of rendering vegetation structures. Relatively simple texture maps can replace complex three-dimensional geometries and microstructures. This allows high visual complexity without excess effort on geometric complexity. Convincing representation of vegetation can also be achieved through a combination of detailed geometry where needed (stem and large twigs) and detailed texture (e.g. small twigs, leafs and flowers) mapped on simple polygons where the geometry is incidental.

Prusinkiewicz and Lindenmayer (1996) developed a different polygon-based approach. Their L-System, which allows the rendering of photorealistic plants, is based on a formal language describing the natural growth of the plants. House *et al.* (1998) applied hierarchical level-of-detail modelling in order to create a very convincing computer-generated walkthrough of an existing forest. Deussen *et al.* (see p. 56) have shown that the complexity of a visual representation of plant ecosystems can be addressed efficiently by combining the use of different levels of abstraction at different stages of the modelling and rendering process (Deussen *et al.* 1998; Deussen 2003).

Animals and humans

For a long time, the visualization of animals and humans in a landscape context was either omitted, or a character animation was the sole purpose (e.g. Magnenat-Thalmann *et al.* 1987, 1989). From a broader ecological view, animals and humans need to be included as they are an important factor influencing human visual perception and shaping the landscape. The presence or absence of animals and humans in the real landscape can greatly influence an evaluation (Hull and McCarthy 1988).

In contrast to essentially static landscape elements such as vegetation, what is especially complicating in the representation of animals and human is their inevitable movement through space. The virtual oceanarium project, for example, permits exploration of several aquatic ecosystems from around the world in a moderate degree of realism, but in three-dimensional stereo projection (see Fröhlich 2000). It is also possible now to encounter virtual humans that deform themselves during motion (Magnenat-Thalmann and Thalmann 2001). These simulated people – often called avatars – can be placed in a simulated environment and can then determine their own actions as independent agents (Farenc *et al.* 2000).

Technologies to capture the complex geometry of the bodies of animals and humans received a strong impulse from computer game development, automotive design, medical and military applications. Nowadays, three-dimensional scanners allow scanning, and hence model building, of the shape and colour of a whole human body at once (e.g. www.cyberware.com). Optical motion tracking systems (e.g. www.ascension-tech.com) can capture up to 900 measurements per second. This may be combined with a stereo three-dimensional projection so users can directly interact with characters embedded within virtual worlds.

Virtual stewardesses now explain safety measures on aircrafts. Movies feature highly realistic virtual actors. Even the subtle expression of moods, age transitions or the gravitational behaviour of hair can be simulated. Walter *et al.* (2001) automatically produce mammalian models with individual bodies at different ages and their associated coat patterns.

Despite these prodigious developments, and although some experimental methods exist for specific environments (Cruz-Neira 2003), we are still some way from simulation and visualization of autonomous animal or human behaviour.

Water

Water is only occasionally static. Apart from a lake surface on a quiet morning, water is a very dynamic landscape element. It takes many forms – rushing streams, waterfalls, waves – which interact in complex ways with the terrain over which they are moving. Fournier and Reeves (1986) modelled ocean waves where the disturbing force is wind and the restoring force is gravity. Even the foam generated by the breakers was modelled by particle systems. This concept is now integrated in standard software running on PCs allowing particles to react with dynamic fields such as gravity, wind and turbulences.

Built structures

In our cultural landscape, built structures play a significant role in creating a sense of place. Integrating built objects in a virtual environment can be a very labour-intensive process. Nowadays, nearly all new architectural proposals are created using computer aided design (CAD). However, from a planning perspective it is equally important to include the surroundings as well. This raises two problems:

- typically these existing nearby buildings are not available digitally;
- CAD systems do not include georeferencing and so fitting new structures into existing models is imprecise.

Nevertheless, some cities, e.g. Basel, Switzerland, already work with detailed three-dimensional models on a citywide scale.

Recent software developments make it easier to build simple three-dimensional models which may be sufficient for many visualization needs. These programs use a library of common structural types (e.g. cuboid, cylinder, trapezium) and can model in a photorealistic way from digital photographs quickly and without extensive three-dimensional skills. Despite the simplified geometry, the models look realistic because of the extracted textures which are mapped on the surfaces.

Beck and Steidler (2001) present a new, efficient approach to record and visualize existing built objects. Their approach is based on a semi-automated generation of three-dimensional objects of the built environment, which allows the fitting of planar structures to a measured set of point clouds. The measurements are taken by an operator, and the structuring of the data is done by the computer.

Another promising alternative for detailed data collection for a small number of buildings is ground-based three-dimensional laser scanning (Manandhar and Shibasaki, 2001). This technology provides a very dense point cloud which can then be structured with specialized software.

Atmosphere and light

The appearance of all these landscape elements can vary greatly under different atmospheric conditions. Influences, including the position and the related intensity of the sun, objects obscuring the light source and general atmospheric or weather conditions, affect the hue, saturation and lightness of all surfaces. Simple fog models reduce the saturation of image pixels based on depth. Clouds can be texture mapped onto a sky dome. However, the atmosphere is often more complex than these simple approaches can effectively portray.

Early in computer graphics history, Reeves and Blau (1985) developed particle systems for the visualization of vegetation elements, water movement, fire, explosions and flocks of birds. Nowadays, several commercial software packages offer particle systems for the visualization of atmospheric effects, such as snow and rain. Based on the OpenGL library freely definable volumetric cloud layers, consisting of several different cloud types, can be rendered in real time, even reacting to wind speed and wind direction (Beck 2003).

The two most popular methods for calculating realistic images are radiosity and ray tracing. Both methods allow computation of the effects of lighting and can output images of photorealistic quality. With ray tracing, scenes that include specular reflections and transparency can be simulated very effectively. For scenes with a high number of light sources and only diffuse reflections, the radiosity approach has advantages: particularly it allows for real-time movement through the model. Kaneda *et al.* (1991) and Nakamae *et al.* (2001) provide good examples of landscapes under natural sunlight conditions.

Perceptual and societal issues

Research in landscape visualization has, to date, focused on technological issues and incredible advances have been made. It may not be long before we see photorealistic interactive virtual worlds where avatars (virtual humans) interact with each other and the environment in natural ways. However, in contrast to the technological challenges, perceptual and societal issues of visualization have hardly been touched (cf. Lange 1999; Bishop *et al.* 2001) in landscape visualization research.

Because landscapes are highly complex, visualization can be incredibly laborious. On the other hand, the omission of details of the real landscape makes for a certain sterility of virtual landscapes (Ervin 2001). How much reality is needed for effective environmental planning? What is the difference in terms of perception between naturalism ('looking like') and realism? Daniel and Meitner (2000: 69) suggest that:

> more abstract representations appear to be inappropriate for determining landscape aesthetic/scenic beauty values. An important question for further research is to determine what representations are necessary and sufficient to achieve valid indications of the effects of particular environmental conditions and characteristics on specified behavioral, perceptual or valuation responses.

Meanwhile, Appleton and Lovett (2003) have undertaken perception studies using commercial software to determine how acceptance of GIS-based visualization varies with level of detail. While studies in landscape have tended to reinforce the importance of realism, Pietsch (2000) has argued the contrary in an architectural, urban design context. This debate continues.

For many years now, digital visualizations have been available for planning. However, they are often regarded as the pretty pictures produced at the end of a linear planning process. The great potential of landscape visualization lies in its early integration in the planning process. Only if the (still pretty) pictures are an integrated and integrating part of the planning process we can expect better and more informed results. The major reasons why this change has not happened already appear to be:

- the lack of user-friendly and intuitive software tools for easy manipulation and design of the landscape (as proposed by ZALF 2002). Visualization has required a high level of specialized skills thereby limiting a widespread application in practice;
- the lack of coupling of visual representation with underlying landscape functions (Hehl-Lange 2001) or political processes which create the specific visual appearance of the landscape;
- the inability, in most cases, of the visualization to support interactive manipulation of design or planning elements.

Some emerging programs (e.g. Kwartler and Bernard 2001) allow alternative land development policies to be formulated as maps, through tables or as the output of scenario models. The alternatives are then directly translated into a three-dimensional model allowing interactive exploration of scenarios, with underlying functional models and minimal technological expertise (see also Chapter 10, p. 251).

Over the next decade, these developments will become widely accessible to people working in spatially relevant disciplines. It is then up to the professionals in landscape and environmental planning to take advantage of the opportunities offered.

CHAPTER 2

VISUALIZATION CLASSIFIED

Ian D. Bishop and Eckart Lange

The distinguishing features of visualization

As Jostein Gaarder (1996: 127) wrote in the popular philosophical novel *Sophie's World*:

> We can even trace a particular word for 'insight' or 'knowledge' from one culture to another all over the Indo-European world. In Sanskrit it is *vidya*. The word is identical to the Greek word *idéa*, which was so important in Plato's philosophy. From Latin we have the word *video*, but on Roman ground this simply meant to see. For us, 'I see' can mean 'I understand'.

This link between seeing and understanding was the basis for the adoption of the term 'visualization' by McCormick *et al.* (1987). This was a new use of the word. Earlier dictionary definitions were restricted to the process of forming a mental image of, or envisioning, something. The more recent usage involves the process of interpreting something in visual terms or, more particularly, putting into visible form. Tufte (1990), for example, provides a marvellous review of mechanisms to help people to envision information. In this chapter we are specifically concerned with the options for putting things (data, model outputs, landscapes) into visual form. The overriding purpose behind creation of these visual forms is to help people to envision. That is, to better understand the relationship between data or some condition of the environment.

More specifically, McCormick *et al.* (1987) defined visualization in this way:

> Visualization is a method of computing. It transforms the symbolic into the geometric, enabling researchers to observe their simulations

and computations. Visualization offers a method for seeing the unseen. It enriches the process of scientific discovery and fosters profound and unexpected insights. In many fields it is already revolutionizing the way scientists do science.

In the very same year (Zube *et al*. 1987) wrote:

> For centuries it appears to have been assumed that a drawing – is a drawing – is a drawing, and that it probably means the same thing to all who view it. The evidence ... suggests that the most realistic simulations, those that have the greatest similitude with the landscapes they represent, provide the most valid and reliable responses (p. 76).

Two different strands of meaning are now emerging. There is visualization of data, models and relationships, and there is visualization of landscape and changing environments. Within these broad categories there are divergent strands. A full understanding of visualization (in its modern meaning) requires close examination of the many options which continue to emerge with evolving technologies and applications.

An early attempt at identification of the major variables in visualization for cartographic purposes was proposed by MacEachren *et al*. (1994). Major axes of distinction were identified as:

- applications for presentation of known information versus the discovery of new knowledge;
- the level of interaction available with the data;
- the use in a private or public context.

Moving beyond the cartographic context, we recognize a number of other important distinctions that can be made within the realm of visualization. These include:

- abstract versus realistic presentation;
- dynamic versus static views;
- single versus multiple displays;
- immersive versus non-immersive display.

All of these differences in visualization options are reviewed in more detail below. With the addition of levels of abstraction, dynamism and immersion into the classification, examples within each division can all be of diverse character themselves. Almost all the examples could be made available in immersive or non-immersive environments.

Visualization classified

Communicating versus discovering knowledge

Maps are a traditional mode of information communication. So are renderings of proposed buildings. Figure 2.1 shows examples of innovative visualization work aimed specifically at communication. Figure 2.1(a) shows the route taken by Lewis and Clark as they opened up what later became the Oregon Trail. The USGS mapmakers have taken great care in communication of known information in an easy-to-read format. The combination of the mapped path of discovery with the elevation-based shading also gives a clear idea of how the path was shaped by the geomorphology of the region. Architects have been rendering images of their designs for several years to show the public how the proposal will look (Figure 2.1(b)).

Discovery of knowledge is clearly the objective in the exploratory, pathfinding work at the USA EPA visualization group. Figure 2.2, for example, shows a combination of three-dimensional representation of the modelling environment with innovative display of the output of a wind velocity modelling program. In the original you can see the combination of colour (for velocity) and icons (for direction) in the wind cross-section.

2.1 Example of visualization for public communication: (a) the line shows the route of Lewis and Clark as they opened the American west; (b) elaborate architectural rendering

Understanding visualization

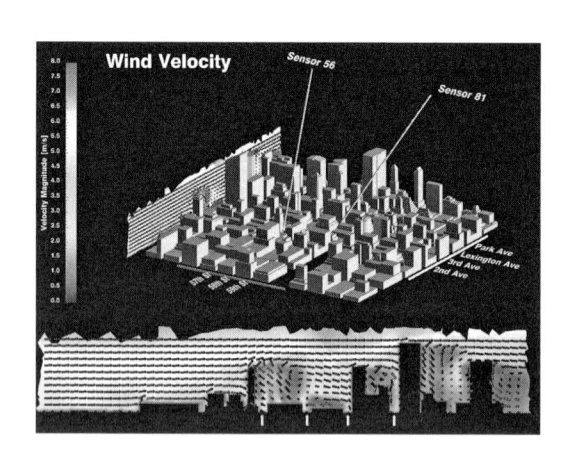

2.2 A prime example of scientific visualization for the purposes of discovery

Static and dynamic display: spatial or temporal

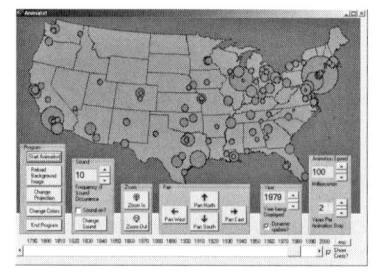

2.3 Interactive and dynamic mapping of population of the 100 largest cities and other urban places in the US 1790–1990

'Dynamic representation' refers to the displays that change continuously, either with or without user intervention (Slocum *et al.* 2001). One form of dynamic representation is an animated map, in which a display changes continuously without any direction from the user. The other form is direct manipulation, which permits users to explore spatial data by interacting with mapped displays (described under 'Levels of interactivity' below).

Many different kinds of changes may be represented by dynamic displays. In the field of cartographic visualization it is temporal phenomena which are most commonly displayed dynamically. This includes the spatial distribution of data on population, pollution, sea surface temperature, mortality rate and so on. While in the early days of development of dynamic visualization it was a novelty to produce an animation, today the user typically has considerable control over functions such as map zoom and pan, changes of colour, sound, projection, animation speed and step size (Figure 2.3).

In realistic visualization, the dynamics are more commonly spatial (i.e. changing view-point) and would normally be in the form of a fly-over or walk-through. Again, this may be a pre-rendered animation or an interactive exploration. However, temporal change is also frequently included in the more realistic three-dimensional style of visualization. In the landscape this may involve the harvesting, burning and growth of trees (see Chapter 6, 'Studying the acceptability of forest management practices using visual simulation of forest regrowth', p. 112), the movement of animals through the landscape (Hehl-Lange 2001), or restoration of a mining environment (Hehl-Lange and Lange 1999).

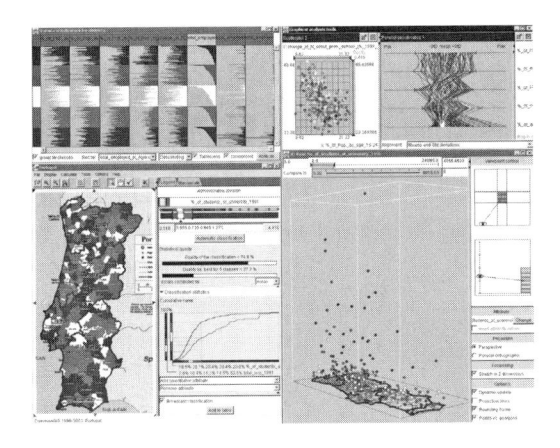

2.4 An example of a web browser accessing an interactive spatial data analyser CommonGIS: an abstract mapping system

Levels of interactivity

The ability to interact with a visualization is the key to its use for discovery, and may also be important in communication.

Some of the classic work by Gennady and Natalia Andrienko (e.g. Andrienko and Andrienko 1999) is illustrated in Figure 2.4. These maps of thematic data, available over the Internet, are highly interactive. The user can move sliders, change a number of display categories, zoom the map and so on. This is an example of visualization which supports both communication and discovery and is available to both the public and the professional.

The degree of interactivity in visualization depends in part on the nature of the controls provided by the user interface. However, another element of interactivity is the ability of the computer systems to redraw images quickly enough that the user sees an 'instant' response to their input. The issue of what constitutes real-time performance and how this is influenced by computer configurations is discussed further in Chapter 4.

Dimensionality

We recognize maps as two-dimensional graphics and rendered buildings as three-dimensional but there are more distinctions that can be made in terms of dimensionality. Particularly in the landscape field, and when working with GIS, some representations are referred to as 2.5D (or two-and-a-half dimensions). This description comes primarily from the way terrain is modelled in many systems – as a single Z value for each X,Y pair in the data set. While this can effectively model most terrain surfaces except caves and overhangs, it cannot be used for full three-dimensional modelling of buildings.

The dimensions of a three-dimensional visual representation do not have to match those of the real world at all. MacEachren *et al.* (2001) have created a distinction based on the extent to which the dimensions of the display environment match the dimensions of the real world. When there is concordance they define the visualization as spatially iconic. In other words, the situation is familiar to us, human perception and cognition are not strained, and real-world metaphors such as digging and flying are readily understood. This is akin to the natural-scene paradigm defined by Robertson (1991). When one of the three dimensions, usually the vertical, is used for something other than the geographic dimension of height, the visualization may be described as spatially semi-iconic. Thus, X and Y may be normal spatial extent but Z is some other variable – population density, income, pollution level. The third option is that the axes of the display environment are quite unrelated to real world dimensions: e.g. various forms of statistical plot.

Levels of realism

Spatially abstract visualization requires a substantial degree of familiarity with both the subject matter and the display technique. Abstract display is therefore generally considered to fall within the expert and knowledge discovery domain. An exception is the familiar static two-dimensional map. However, times change and within the spatially iconic communications domain different levels of detail, realism and abstraction may also be appropriate to permit ready public consumption.

Part of the role of visualization for the public is to provide an opportunity for greater involvement in community decision making. Government, or consultant, reports are often designed for people with an existing knowledge of the issues or processes involved. To broaden the effective use of this information it needs to be in a format (or language) that can be widely consumed. Haughton (1999) has warned of the dangers of 'technocratic capture of information by dint of its poor presentation for interpretation by ordinary citizens' (p. 54). Environmental management now covers a very wide range of issues and the public are concerned with a great many of these. At the same time as technology has advanced the opportunity for visualization, public interest has increased the need.

Langendorf (2001) makes four assumptions about the role of visualization in planning:

1. in our complex world, to understand nearly any subject of consequence it is necessary to consider it from multiple viewpoints, using a variety of information;

2   we are rapidly moving from an information-poor to an information-rich society;
3   the understanding of complex information may be greatly extended if visualized; and
4   problem solving and commitment to action in a complex world requires communication and collaboration among many participants, and visualization aids this interaction (p. 309).

The easiest form of visualization for the public to associate with and understand is realistic portrayal of visual landscape change. Bishop (1994), Lange (1994) and others have argued that realism is of great importance in effective and legitimate communication of change. Realistic presentation of alternative futures is clearly appropriate when immediate aesthetic issues are uppermost in people's concerns but this form of presentation can also be effective for broader issues. Communication of flood risk, traffic volumes, forest succession (Figure 2.5), visual pollution levels and other factors influenced by environmental management can all be represented with high levels of realism. The approach to communication through realistic images of a real or proposed environment is often referred to as Environmental Visualization (encompassing visual simulation of both built and natural environmental features).

There are clearly environmental management impacts which either cannot be represented realistically (non-visual pollution, regions of influence) or are more easily interpreted by a more schematic form. Figure 2.6 is an example from a landscape planning project where realism was not attempted (see also Chapter 7, 'Designing, visualizing and evaluating sustainable agricultural landscapes', p. 136). The important point was that the geometry was based entirely on existing or proposed terrain and surface features derived from a GIS. Krause (2001) also used a semi-realistic approach but added further information using entirely abstract icons (Figure 2.7). Coupling realistic and abstract landscape representation, Hehl-Lange (2001) uses the real terrain as the visual reference for the abstract representation of barriers to move-

2.5 Frame from an animated sequence of forest regrowth after harvest. The sequence presents more than just immediate visual consequences and gives viewers a clear understanding of rates of growth and species mix and density

Understanding visualization

2.6 Three-dimensional landscape representation but with solid colours chosen to match those in corresponding GIS mapping (image reproduced in the colour plate section)

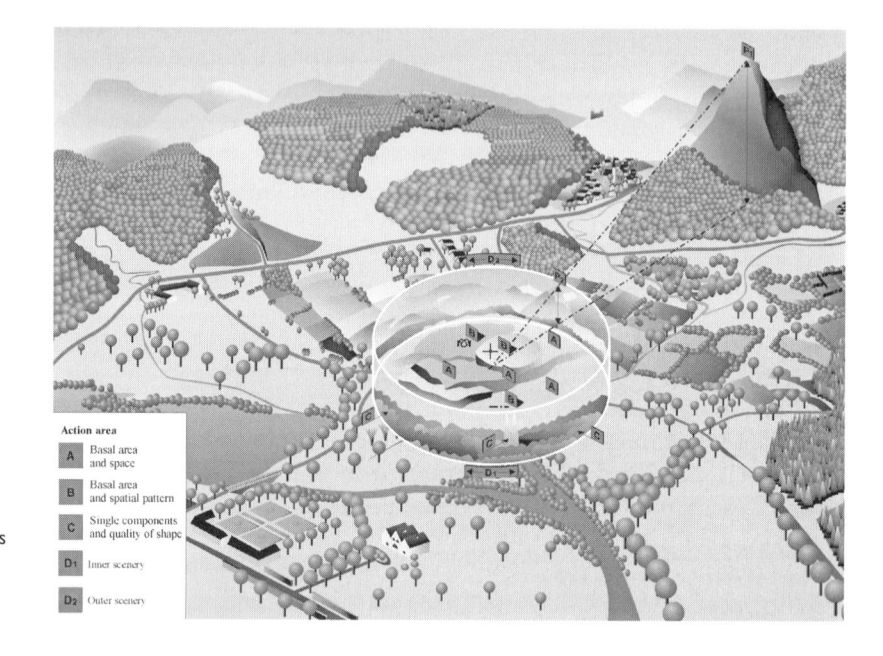

2.7 Here the semi-realistic representation of landscape elements is supplemented with additional symbols to communicate information about proposed landscape change

2.8 A realistic representation of the base terrain creates a context for the abstract information on the accumulated barrier effect to movement of the common toad

ment of the common toad in central Switzerland (Figure 2.8). There are many examples, but little research has been undertaken to determine what works best in each context.

In considering realism in landscape portrayal, a useful distinction can also be made between surface textures based on actual photographs (aerial or ground level) of the location being modelled (*geospecific* textures) or textures which are typical of the ground covers, building facades or tree types of the region (*geotypical* textures). These terms were introduced by Graf *et al.* (1994). Geospecific texture can be expected to create a more realistic model, but geotypical textures give greater flexibility in that changes in landcover over time can be easily modelled. A similar distinction can also be made in terms of the geometry of objects. Tree models, for example, are almost always geotypical. We seldom attempt to reproduce an individual tree specimen. Building geometry can, however, be specific or typical. Perrin *et al.* (2001), for example, use geotypical building geometry and textures for their GIS-driven visualizations (Figure 2.9(b))

2.9 The distinction between geospecific and geotypical textures and objects: (a) geospecific ground texture and geotypical trees; (b) geotypical ground textures and housing (images reproduced in the colour plate section)

Understanding visualization

Single versus multiple representations

Just as any of a number of different views of data can be represented in a single image, so we can also take the option of showing multiple views simultaneously or give the user the option of moving rapidly between alternative views; whether they be abstract or realistic, static or dynamic, etc. As we shall see in the applications-oriented chapters in Part Two, there is a role for many different styles of visual representation according to the application and the audience. In many circumstances it is desirable, or essential, to have more than one form of visualization available such that:

a   one user can have multiple views; or
b   that multiple users can chose the view that best suits their experience or professional background and information needs.

When working in an interactive environment, a factor that also then becomes important is which of the views is available for interaction. Is one view always preferred for interaction? In GIS products, for example, the user may view maps, tables, charts and 2.5D perspectives but can only edit the data via the map (often a complex process) or the attribute tables. Batara *et al.* (2001) argue that, ideally, the urban designer, for example, should have the option to interact with any of the representations as they relate to different aspects of the design process from overall financial outcomes to aesthetics. Any changes in one view must, of course, be transferred to the database and hence to the other representations.

The developments described first by Germs *et al.* (1999) and Verbree *et al.* (1999), which later became part of a commercial visualization system, emphasized the concept of multiple data views and referred to three main options as plan view, model view and world view (Figure 2.10).

2.10 The three views of GIS data as defined by Verbree *et al.* (1999): (top) plan view; (middle) model view; (bottom) world view (images reproduced in the colour plate section)

The concept of virtuality: virtual reality and virtual environments

Recently, a lot of attention has been paid to the concept of virtuality in visualization. The thinking here is that there is a special class of visualization which is sufficiently distinctive to be called 'virtual' – as in virtual reality or virtual environment. The original reason for using the term was that one could be made to feel virtually (i.e. almost) present in some artificial environment. More recently, 'virtual' has been used more loosely to encompass any computer-based interactive model of reality – especially if the user can run physical process models and

Visualization classified

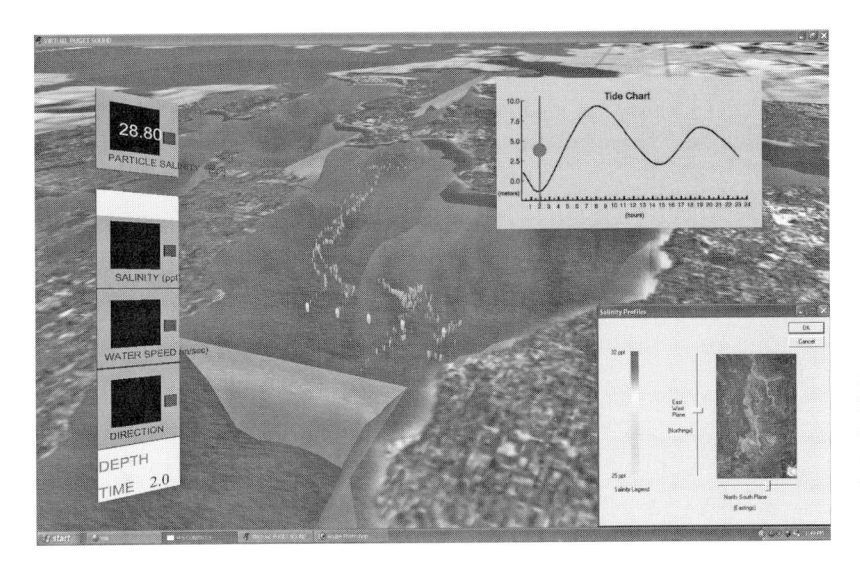

2.11 Virtual Puget Sound (Winn et al. 2002 – within the fully immersive virtual environment (HMD or CAVE) special interface procedures may need to be developed. Shown are the gesture recognition navigation tool, a high-level menu and the tide cycle controller

undertake 'what if' type experiments within the modelled environment (Figure 2.11).

There are a number of features of computer-generated environments which many authors consider important, perhaps essential, for qualification as 'virtual'. Jacobson (1993) defined VR as an 'environment created by the computer in which the user feels present'. About the same time, Sherman and Judkins (1992) proposed five key factors as the basis of virtual reality. Later, Heim (1998) proposed a three-factor model, then MacEachren et al. (2001) used this as a base and added a fourth factor. Table 2.1 summarizes the available factor definitions (which the original authors all began with the letter I). In our view, the key factors are the first three – immersion, interactivity and realism.

Sherman and Craig (2003) separate the Interactivity criteria into two parts which they call sensory feedback (body tracking) and interactivity (object manipulation). More importantly, they separate the virtual world (the collection of virtual objects and the rules which govern their behaviour) from the medium through which the virtual world is accessed. This separation of the world from the medium also permits a clear definition of a *virtual environment* as being:

1 a virtual world
2 an instance of a virtual world presented in an interactive medium such as virtual reality.

This clarifies the situation in landscape visualization where we often use very detailed data to represent the virtual world but present this

Understanding visualization

Table 2.1 Features deemed significant to a virtual environment

Feature	Definition
Immersion	Immersion describes the sensation of 'being in' the environment (Heim 1998 and MacEachren *et al.* 2001)
	VR should deeply involve or absorb the user (Sherman and Judkins 1992)
Interaction	Enables a participant in a virtual experience to change their viewpoint on the environment and to change the relative position of their body (or body parts – hands) in relation to that of other objects (Heim 1998)
	Enables manipulation of the characteristics of environment components (MacEachren *et al.* 2001)
	In VR, user and computer act reciprocally through the interface (Sherman and Judkins 1992)
Intensity (realism)	The detail with which objects and features of the environment are represented (Heim 1998 and MacEachren *et al.* 2001)
	In VR the user encounters complex information and responds (Sherman and Judkins 1992)
Intelligence	The extent to which components of the environment exhibit context-sensitive 'behaviours' that can be characterized as exhibiting 'intelligence' (MacEachren *et al.* 2001)
Illustration	VR offers information in a clear, descriptive and illuminating way (Sherman and Judkins 1992)
Intuition	Virtual information is easily perceived. Virtual tools are used in a 'human' way (Sherman and Judkins 1992)

world through simple static images or animations because the detail is too great for real-time virtual-reality based presentation. The trade-off between detail and interactivity is still very real in visualization of landscapes.

CHAPTER 3

DATA SOURCES FOR THREE-DIMENSIONAL MODELS

Ben Discoe

Basic concepts

To create a model of a landscape, the first steps are to learn about the data sources and to acquire geospatial data for the area of interest.

There are a nearly infinite number of ways in which the human mind can subjectively conceptualize a model of the real world, and each results in a different form of geospatial data. To cover some of the most general ways, we can consider the following broad categories: elevation, imagery, vegetation, water, buildings and transportation.

Much of the terminology of landscape visualization has developed over the past few decades in the fields of remote sensing and 'visual simulation', or Vis-sim, which is the name used by the military and aerospace industry to describe the process of creating terrain representations for training and testing purposes. This terminology includes the following:

- all features that are placed on the ground are called *culture*, including buildings, roads, and vegetation;
- imagery is classified into two kinds, *geospecific*, meaning an image which explicitly describes a particular location, and *geotypical*, which is a generic image of a kind of ground, such as grass or sand;
- data whose geographical location is known to some degree of accuracy is known as *georeferenced*.

Often, when searching for data suitable for landscape visualization for an area, there may be no ready-made sources of digitized data. In some cases, digitized data may be available but at the wrong scale. The best available data may be conventional paper maps, such as topographical maps or nautical charts.

Understanding visualization

However, the range of available digital data is expanding rapidly. A list of sources for free and licensable geospatial data would be huge and rapidly obsolete due to the vastness of the field and rapid pace of change. There are many online directories including the Virtual Terrain Project (vterrain.org) and The GeoCommunity (geocomm.com) which can be consulted. In each country (or state, or province) there is increasingly likely to be a government agency providing a portal to national (or regional) digital data sources. When sources of spatial data are linked to each other or to a central server, it is now commonly referred to as *spatial data infrastructure* (SDI).

Working with conventional maps

The first steps in making use of paper maps are to:

- determine the map's projection and extents – the coordinates at the edges of the maps, hopefully printed or derivable from the map;
- digitize the map with a scanner, producing a high-resolution, high-contrast image.

For a topographical map, the next step is to extract the contours and other desired linear features, a process known as *raster to vector conversion*. The scanned map (*raster*) can be processed by software packages capable of semi-automatically resolving the linear features (*vectors*). Generally, there are still manual steps involved in correcting the automatic results, such as fixing gaps and assigning an elevation value to each contour.

For a nautical chart, there may be contours for the shoreline and near-shore areas, but the underwater elevation is most likely to be represented by loosely-spaced depth values. Digitizing these points is a manual process. Turning the points into a form useful for visualization, such as a grid or a TIN, involves a process of interpolation. This can range from a naïve triangulation to an advanced process of weighting each input, called *kriging*. Producing good results with kriging is an art as well as a science. The result will be an elevation surface which can be visualized the same way as regular dry land.

The remainder of this chapter deals with digital data sources.

Coordinate systems and projections

It is vitally important, when dealing with geospatial data, to know and respect the coordinate reference system (CRS) in which the data is defined. All commonly-encountered CRS are based on *geographic*

Data sources for three-dimensional models

coordinates, commonly referred to as latitude and longitude or *lat-lon*, which use units of degrees. The data may be projected into linear units using any of several *projections*, at which point it is referred to as a *projected coordinate system*. Without a projection, the CRS is simply a *geographic coordinate system*.

The most common projections are listed below.

- Transverse Mercator (TM) which projects the earth onto a cylinder which touches the surface at a given longitudinal line. It is well suited for areas which are more tall (north–south extent) than wide.
- Universal Transverse Mercator (UTM) is a set of 60 standard TM projections, known as *UTM Zones* (Figure 3.1).
- Albers Equal Area (AEA) and Lambert Conformal Conic (LCC) are popular projections for data which is large in both height and width.

Underlying any CRS is the concept of a *geographical datum*. A datum is a convention for mapping geographic coordinates onto the real earth, and there are many systems that have been used historically. Fortunately, most modern data will use the latest internationally standard datum, the *World Geodetic System of 1984* (WGS84). Some commonly encountered older datums include the following.

- NAD27 and NAD83 (North American Datum). NAD83 is functionally identical to WGS84 over the continent of North America.

3.1 The UTM zones of the world (from http://www.dmap.co.uk/utmworld.htm)

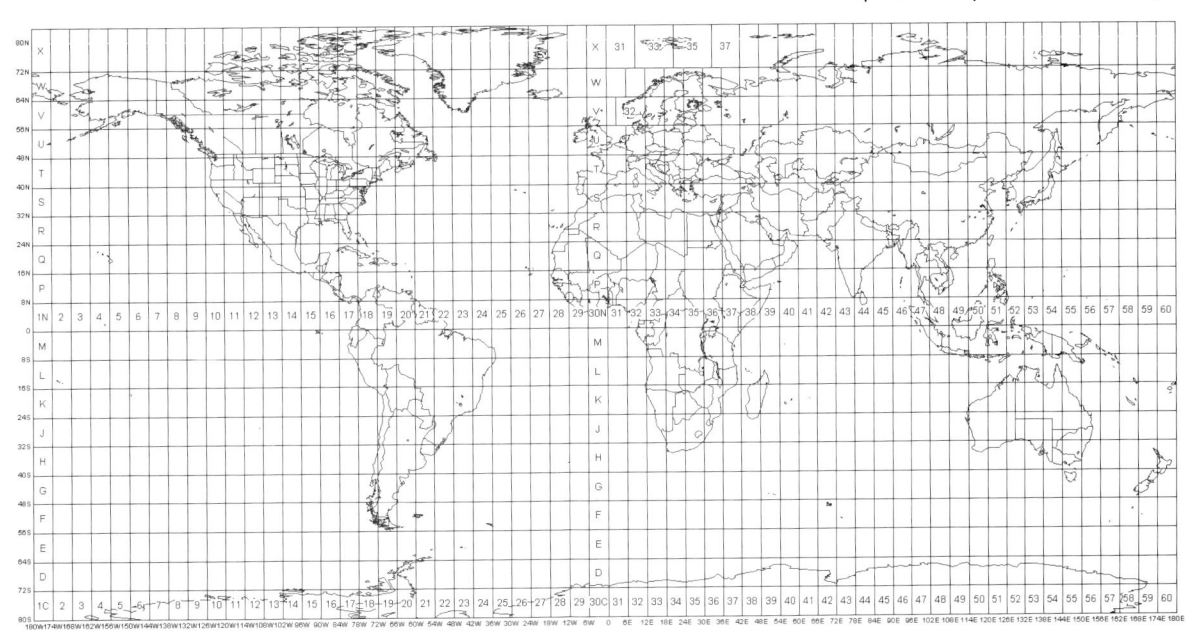

37

Understanding visualization

- WGS72, a previous version of WGS84.
- European Datum 1950, the UK's OSGB36, and many other datum in Europe.

Another phrase for 'coordinate system' is 'spatial reference system' (SRS). The abbreviation SRS appears often in standards documents, such as those of the Open GIS Consortium (OGC).

There are several well-known ways in which to represent a CRS. The oldest is a set of codes defined in the 1980s by a group called the European Petroleum Survey Group (EPSG). They introduced a numbering system for all known datums and projections, and a set of the most commonly used CRS. These codes are popular because a single small number is used. For example, '32605' represents 'the projection UTM Zone 5 Northern Hemisphere based on the WGS84 Datum'.

A more verbose way to express a CRS is the OGC standard WTK (Well-Known Text) representation. For example, the same CRS would be expressed like this:

PROJCS["UTM Zone 5, Northern Hemisphere", GEOGCS["WGS 84", DATUM["WGS_1984", SPHEROID["WGS 84", 6378137, 298.257223563, AUTHORITY["EPSG","7030"]], TOWGS84[0,0,0,0,0,0,0], AUTHORITY["EPSG","6326"]], PRIMEM["Greenwich",0, AUTHORITY["EPSG","8901"]], UNIT["degree",0.0174532925199433, AUTHORITY["EPSG","9108"]], AXIS["Lat",NORTH], AXIS["Long",EAST], AUTHORITY["EPSG","4326"]], PROJECTION["Transverse_Mercator"], PARAMETER["latitude_of_origin",0], PARAMETER["central_meridian",-53], PARAMETER["scale_factor",0.9996], PARAMETER["false_easting",500000], PARAMETER["false_northing",0]]

As you can see, this takes much more space, but its flexibility allows you to express any coordinate system. For example, the commonly used 'State Plane' CRS in the USA, which use feet instead of metres as their units, are supported by WKT, but not by EPSG codes.

Elevation data

Elevation is represented in many diverse ways, including grids, points, contours and triangular irregular networks (TIN) (Figure 3.2).

Gridded elevation data is the most common and the easiest to work with. The USGS, the largest source of publicly available elevation data in the world, uses the *DEM format* to distribute elevation grids for the United States. The file extension is *.dem* and it is a reasonably straightforward, though inefficient, text format. DEM is widely supported by software packages for reading, but not for writing, as it is not well suited for use as a data exchange format.

Data sources for three-dimensional models

3.2 The same area represented as: (top left) point elevations; (top right) contours; (bottom left) TIN; and (bottom right) grid

Some other elevation grid formats include:

- DTED, a text format used by the US military. It comes in several resolutions, from DTED0 (30 arc seconds/ ~1 km), which is publicly available, to DTED1 (3 arc seconds/ ~100 m) and DTED2 (1 arc second/ ~30 m) which are not;
- GTOPO30, a format used for a popular, freely-available 1 km global dataset;
- CDF and HDF are formats common in the scientific visualization field, often used for underwater elevation (*bathymetry*);
- ArcInfo Binary Grid Format, a common yet proprietary format, can be read by many non-proprietary software tools;
- BT (Binary Terrain) is a compact format intended for data exchange.

Elevation point data is often encountered with bathymetry, since methods for surveying the surface of underwater areas have historically been limited to sampling individual depth values. Unfortunately, there are no standard formats for this form of data.

Contours are widely used to represent elevation, especially with historical maps (*topographical* maps, or *topo* maps) and for surveying of small areas such as individual parcels of land. A contour line is simply a line of equal elevation, which turns and twists to follow the contours of the terrain. The technical term for using contour lines for elevation is *hypsography*. Although they are easy to understand, and widely used in surveying, contours have many drawbacks and limitations. Significantly, contours are not efficient to render directly for visualization. For this reason, elevation represented as contours is generally converted to a grid or a TIN in order to visualize it.

It is also difficult for a contour representation to model a sharp change in elevation or an area of fixed elevation. A *breakline* is a mechanism for overcoming this limitation, by explicitly indicating elevation along a line. If breaklines are present, they are very valuable for producing an accurate conversion to a non-contour representation.

A TIN represents an elevation surface as a network of triangles, which can be any shape or size but generally covers the ground without gaps or holes. A TIN has advantages over a grid in that it can precisely model changes in elevation, such as the curved edges of a roadway or the ridges of a mountain, without being limited to a fixed resolution. A TIN also has many drawbacks, including requiring much more storage space (XYZ must be stored for each point, and the connectivity information, instead of simply elevation) and it is slower and more complicated for operations such as height sampling.

Image data

Imagery, generally captured from a satellite or airplane, comes in many varieties, depending on the purpose for which it was captured. There are many wavelengths of light, such as the infrared which are important for analysis in remote sensing, and these can play an important role in landscape visualization as well. Some of the terminology is explained below.

- *Multispectral* images are composed of one or more *bands* each containing the image at a certain range of wavelengths.
- Common *bands* include those of visible light, generally 'red', 'green', 'blue', and other non-visible wavelengths such as 'near infrared'.
- *Panchromatic* images have a single value, generally an average of several bands. They appear monochromatic (greyscale).

Data sources for three-dimensional models

Aerial images (from an airplane) are usually either visible light or infrared. Satellite images are usually only partly visible light, since the visible spectrum (especially blue) does not travel clearly into space, for the same reason (scattering) that the sky looks blue to us.

Satellite images

Some of the most widely known sources of satellite imagery include *SPOT* (France), *IRS* (India) and *LandSat* (USA). LandSat is a series of satellites, the most recent being LandSat 7 (1999–present). The bands captured by LandSat 7 are typical:

1	0.45–0.52 µm	Blue-green
2	0.52–0.60 µm	Green
3	0.63–0.69 µm	Red
4	0.76–0.90 µm	Near IR
5	1.55–1.75 µm	Mid-IR
6	10.40–12.50 µm	Thermal IR
7	2.08–2.35 µm	Mid-IR

Since the blue band is generally too fuzzy, a common approach is to use green for blue, and one of the infrared bands for green. This is an example of *false colour* imaging, since the colours of the resulting image are artificial, rather than the true frequencies of the original image. This particular approach sometimes produces reasonable results since infrared is reflected most strongly by green vegetation. However, this also produces unnatural colour results, such as cities and deserts appearing purple. Until recently, satellite imagery had too low a resolution (Landsat 30 m, SPOT 10 m) to be effective for simulation of small areas. New sensors like Ikonos (1999) and Quickbird (2001) provide images with a resolution as small as 60 cm.

Aerial images

Aerial photography has a long history of greyscale ('black and white') imagery, which has traditionally been used for purposes including stereographic tracing of elevation contours and land classification for agricultural purposes. Colour imagery is a more recent phenomenon.

One issue to be aware of is that, for many landscape visualization purposes, aerial images can actually be *too* high resolution. A feature like a tree or a building, when draped at high resolution on elevation data, looks quite wrong flattened onto the ground. Although sub-meter imagery can look very impressive, it does not avoid the need for three-dimensional modelled culture.

Image file formats

Most traditional images file formats, such as JPEG, are also used for geospecific images, but several formats are favoured for particular use in this field:

Understanding visualization

- GeoTIFF, a flavour of the common TIFF format (.tif) which embeds the geographical CRS and location (georeferencing) of the image in its header;
- MrSID (.sid) and ECW (.ecw) are proprietary image formats which are highly compressed, can store georeferencing, and can be quickly uncompressed at desired levels of detail.

Traditional image file formats such as JPEG can be extended to include georeferencing information by means of a *world file*, which is a small text file containing coordinate information for the image. For example, TIFF files created before the advent of GeoTIFF may include a *TIFF world file* (.tfw). You may also encounter a *projection file* (.prj) which contains a description of the CRS of the image.

Surface textures

At visualization time, the colour of the ground will generally be represented by a *texture map*. The terminology comes from computer graphics and does not actually mean 'texture', but rather colour. This texture map can come from one of the satellite or aerial images discussed above, which describes a specific area of the Earth's surface, in which case it is known as *geospecific*. Alternately, a set of representative textures such as sand, rocks and grass can be combined to produce an artificial texture. In this case, the source textures are known as *geotypical*, since they are typical of a particular kind of surface (see also Figure 2.9 and accompanying text).

These approaches are not mutually exclusive – sophisticated landscape visualization software can use a geospecific texture from a far view distance, then replace it with best-guess geotypical textures when the viewpoint is close to the ground.

There are pitfalls in this approach to watch out for. For example, consider the case of a green tree. From a distance, in a top-down geospecific

3.3 In the foreground, both tree objects and trees in the geospecific texture image coincide sufficiently well that the trees in the image appear to be shadows of the tree objects. This is fine until changes in tree distribution are modelled

image, it appears as a green speck or, in a high-resolution aerial photo, as a green circle. If this geospecific texture is naïvely applied on the ground, the result looks good from a distance but, on closer inspection, the tree looks simply like a green circle on the ground. If the landscape data includes the location of the tree, it can be drawn as three-dimensional geometry when close, and not drawn from far away. At the transition point, the ground texture must change from a green circle to a best-guess geotypical texture of the ground under the tree. Figure 3.3 shows a geospecific ground texture including trees and tree objects placed at the corresponding location.

Vegetation

Data describing vegetation comes in a very wide variety of forms:

- point features may be used to describe the locations of individual plants, sometimes including attributes for each instance;
- polygonal data (*coverage*) may represent any aspect of the vegetation in that area – species, species mix, height, density, etc.
- bitmap or raster data (*bitmap/raster coverage*) is an alternative to polygons, which also may indicate any number of vegetation attributes.

This is a field traditionally handled by GIS software, so the data is likely to be found in any of the common GIS formats or, for US data, USGS formats. Unfortunately, to date, there are no standards for the attributes in these formats. Attributes vary widely and depend heavily on the particular needs and interests of the producer of the data.

It is sometimes necessary to combine more than one source of data in order to determine vegetation. For example, an agricultural department may provide a *land cover* map, with classification of land into urban, agricultural, desert, sparse and heavy wild vegetation. Biologists may provide a species distribution map, showing the mix of species (sometimes called an *ecotype*, *ecotope*, *ecosystem* or *biotype*). By combining these two sources (data fusion), it is possible to estimate both species and density for visualization purposes.

Other possible sources include using elevation or slope (from an underlying elevation layer) as an input to the distribution heuristic. This is very valuable if certain species are known to prefer specific ranges of elevation or slope, especially in the absence of precise land cover data.

Some types of vegetation such as large trees are well suited to point data, whereas others such as grasses can only be practically represented as a coverage distribution. In fact, many groundcovers are represented as a colony of clones or even a single organism/object with a wide horizontal extent! Efficiently modelling of grasses is still an active research area (see Chapter 4, the section 'Efficient modelling and rendering of landscapes', p. 56).

When rendering vegetation for visualization, there are a few approaches for creating three-dimensional geometry.

- *Billboards* – a texture map with a complete image of the plant is applied to either a single rectangular polygon which turns to face the user, or a set of two interpenetrating rectangles which gives some illusion of parallax.
- *Directional billboards* – a set of texture maps of the plant from several angles is stored, then the appropriate image is used at rendering time. This requires a great deal more memory than simple billboards, but looks more realistic and is still fast to render.
- *Explicit modelling* – a conventional three-dimensional modelling tool, or a software package that knows about plant geometry, can be used to produce a detailed three-dimensional model of the plant. This requires a great deal of memory and is slow to render, but can be worthwhile when the visualization requires specific plant instances to be draw realistically from all viewpoints.

One very active area of vegetation modelling is forestry (see Chapter 6). Modern forestry depends on accurate GIS data for its operation, so there is a large body of work on this subject. Many of the attributes used for representing forests, such as *diameter at breast height* (DBH), a measure of the thickness of the trunk of a tree, evolved specifically for their importance to the forest industry. Unfortunately, there is not a standard for these attributes either, so it is a manual process of discovery when a (typically GIS format) file is found with forest data, to understand and utilize the attribute fields.

Water

Moving and stationary bodies of water are represented in a variety of ways depending on their area of application. The general term for mapping and analysing bodies of water is *hydrography*. The term *bathymetry* is also used, most commonly when describing oceans.

The freely available USGS Hydrography DLG dataset demonstrates a common approach from the cartographic domain: bodies of water are describes as polygons, and rivers are described by polylines (i.e. lines defined by multiple points). This simple approach is sufficient for drawing a map, and for some visualization tasks, but not sufficient for more advanced tasks such as analysis, simulation or navigation. Hence, more detailed models are used.

Historically, the depth of water bodies was first measured as *sample points*, known as *soundings* or *depth readings*. These were gathered because of their importance for navigation of ships. Later, for areas which were well surveyed, contour lines came into use, similar to those for elevation of dry land. Finally, more recently, grid representations have been used, especially for large bodies of water, including *oceanography*.

This progression is fortunate for the field of visualization, since grids are well suited for that task. However, when you are faced with a visualization project for an area that includes water, you are very likely to encounter either contours or, most probably, sample points. These will need to be converted to a grid with appropriate software which performs the interpolation guesswork. Water depth is clearly significant when seeking to simulate underwater conditions. However, it is also relevant to above surface visualization because of changes in water appearance with depth.

Buildings and other artificial structures

There are two common domains for representing buildings: CAD and GIS. Traditional drafting and CAD describes a building with a blueprint drawing, which provides just enough detail to describe how the building is to be constructed, but generally lacks any kind of context such as geographic location. Three-dimensional CAD extends blueprints to an actual three-dimensional model, but still lacks geographic awareness.

The GIS approach, and paper-based cartography before it, describes a building with, at most, a footprint, but it does explicitly state where the building is geographically. A footprint is simply a polygon which roughly describes the area covered by the building, and possibly some overall attribute fields, such as height.

Visualization needs *both* these domains – it needs to know what the building looks like, and where to place it on the terrain. The convergence of these representations is still an area of research and development, but here are two ways in which visualization generally works.

First, in the case where only footprints are known, some software can create *procedural* geometry for the building. This process takes very minimal inputs (such as the footprint and height) and produces a 'best guess' three-dimensional building for rendering. This is sometimes called *parametric representation* because the building is described implicitly by a set of *parameters*, rather than explicitly. An example is shown in Figure 2.9(b). Advantages of procedural representation include efficiency, since very little information needs to be stored, and power, since major changes like adding a storey can be done with one simple operation. The disadvantage is that the uniform approach makes it difficult to represent complex and irregular buildings, which are increasingly common in modern architecture.

The second approach is to leverage the power of three-dimensional modelling software by using existing detailed building models or creat-

Understanding visualization

3.4 An example of a procedural fence from VTP

ing new ones. In this case, only a point dataset needs to be created which describes the location (and possibly the orientation and scale) of each building, with a reference to the file that contains the explicit three-dimensional model.

Types of structures other than buildings, such as fences and walls, are even better suited to procedural representation, since they are very efficiently represented by linear features with attributes, and it would take a lot of time and energy to create an explicit model for each section of a fence in a large area, without significant benefit (Figure 3.4).

Transportation – roads, rails and trails

There are three broad levels of detail at which transportation systems are represented:

1. high-level transit simulation: 'macroscopic', with abstract lines of flow through a network, used for regional planning;
2. discrete transit simulation: 'microscopic', individual vehicles moving along idealized roadways;
3. full-detail simulation: just one or a very small number of vehicles on an extremely physically accurate roadway, used for safety and crash analysis.

Unfortunately, there are not yet any standard file formats, or standards for representation by features and attributes, for any of these levels.

The second level of detail is the one that most concerns landscape visualization. In a GIS, each road is generally represented by a *centreline*, which is an imaginary line down the middle of each road (or lane of traffic). These centrelines can be represented either with polylines, or parametric curves. A transportation file may include topology, meaning the roads and intersections are aware of how they are connected, or it may be a jumbled set of features without topological relationships.

Data will most commonly come from a traditional GIS format, or possibly a distribution format from a mapping agency such as:

- the DLG format used by the USGS, or the newer SDTS encoding of DLG;
- the TIGER format used by the US Census Bureau;
- VPF, the Vector Product Format used by the US military.

In more specialized transportation or road design software, there are some proprietary and research formats including:

- Paramics, a widely used commercial tool, that has its own file formats, and
- EDF, the Environment Description Framework,

which are designed with simulation in mind.

These are, unfortunately, rarely encountered. Most often there is only a single centreline per road, often digitized from an old paper map. Software processing is needed to connect these centrelines into a topology, supply attributes for the roads, define the layout of intersections, and other steps which are needed for either simulation or visualization.

Rendering of roadways for three-dimensional visualization can be accomplished in several ways, depending on the approach to modelling of the terrain.

- *Draping* – the roadways are converted into thin geometry which is draped on top of the terrain. There is a gap between the road and terrain underneath is, which can be hidden with *skirt* geometry, or can be ignored if the viewpoint in the visualization will remain at a reasonable height above the road. The advantage of this approach is that it is fast and simple, and is independent of the method used to represent or render the terrain underneath.
- *Merging* – also called embedding or stitching, in which the terrain surface itself is modified to match the road surface, though a process of 'cut and fill' operations similar to those of real roadways in the domain of civil engineering. The advantage of this approach is a realistic-looking roadway.
- *Carving* – a hybrid approach in which the terrain is first modified by moving its vertices up or down to better fit the road surface.

Understanding visualization

3.5 Carving a terrain to match a road surface: (left) naïve drape with intersection problems; (middle) terrain carved to match road surface; (right) a drape with better results

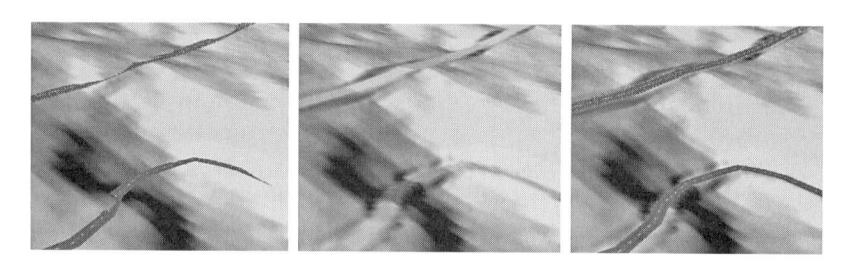

The road geometry is then draped on the terrain, with better results than simple draping (Figure 3.5).

Conclusion

To summarize, the basic steps in creating a three-dimensional visualization of a landscape are to acquire raw geospatial data, process them into an appropriate form, then use them as inputs to software which will construct the three-dimensional geometry (see also Chapter 4).

There are many additional inputs which can be valuable. Direct knowledge of the place being modelled can be immensely important. This can take the form of reference photographs shot from the perspective of the desired visualization, first-hand experience of visiting the location, or the input of a local inhabitant. In fact, some visualizations,

3.6 Wide angle views inside the Palaestra, ancient Olympia: (below) all that is left today; (right) reconstructed from multiple data sources

especially of archaeological sites, sometimes have only reference photos to work from, while others (Figure 3.6) may be based on ancient maps, first-hand descriptions and photographs of remnants (Ogleby and Kenderdine 2001).

In short, nearly every form of recorded information can be valuable for producing a visualization, even such seemingly disparate inputs as oral histories, tree rings, traffic metering, ice cores and cell-phone tracking logs. The act of modelling the real world is as broad and complex as the world itself. Integrating these inputs makes this field well suited to the generalist, while also relying on the specialist for each sub-domain.

CHAPTER 4

VISUALIZATION TECHNOLOGY

INTRODUCTION

Ian D. Bishop and Eckart Lange

Development happens fast in computer hardware – the famous Moore's law suggests a doubling of capability every 18 months. At the time of writing there was opinion that the trend in graphics performance was an astonishing six-month doubling time. We have endeavoured, therefore, to provide information here that will retain currency because it refers to hardware or software issues that remain the same irrespective of speed of performance.

The chapter first reviews the significance of hardware to computer graphics performance. Oliver Deussen and colleagues then review the role of software design in effective environmental visualization, while Adrian Herwig and others explore the potential of game engines for landscape design and decision making. Finally, we review the ways in which the wonderful imagery we are now able to create can be displayed.

In reviewing these technological options, we have tried to give not only a sense of the technology itself but also how it has been used by researchers or practitioners in the context of landscape and environmental planning.

The significance of computer power

Rendering is the process of taking a three-dimensional model, defined by the location of surfaces in space and their surface material, and

Understanding visualization

creating images of those surfaces under specified lighting conditions. The process was pioneered in research laboratories in the 1960s and made its way gradually into modelling, animation, CAD and related software products (see also Chapter 1). One of the distinguishing characteristics of visualizations, as identified in Chapter 2, is the degree to which the user can interact with the display. This is a function of the speed with which pictures can be rendered. If pictures can be created very fast then the user can interact with the presented data or scene in real time, i.e. user action produces a new picture almost instantaneously. Rendering speed is normally measured as the number of pictures the computer can produce per second (frames per second – fps) and is then called the frame rate. It is widely agreed that 30 fps is real time and that 2 fps is not. Between these rates, whether a display feels interactive depends on how fast any camera movement needs to be and what kinds of interactions are being supported.

The achieved frame-rate depends on the power of the computer, the complexity of the modelled world and the sophistication of the software. In this section the basis of computer power (the various elements that can contribute to the frame rate) is reviewed. The discussion is based on the visualization of three-dimensional virtual environments.

Until recently, the performance of a computer was considered primarily in terms of its central processing unit (CPU), the faster and more efficient this was the better. Over the years CPU performance has increased at a remarkably steady rate as exemplified by Moore's law. An engineer from the computer chip maker Fairchild Semiconductor, Gordon Moore in 1965 published a paper (Moore 1965) in which he found, using data since 1959, that the chip content had been doubling each year. He predicted this could go on for another 10 years. Development has been tracked since and the trend still continues although the doubling time is now usually reckoned as about 18 to 24 months (Figure 4.1).

However, CPU power is not everything – especially when it comes to real-time visualization. Since 1992, with the release by Silicon Graphics

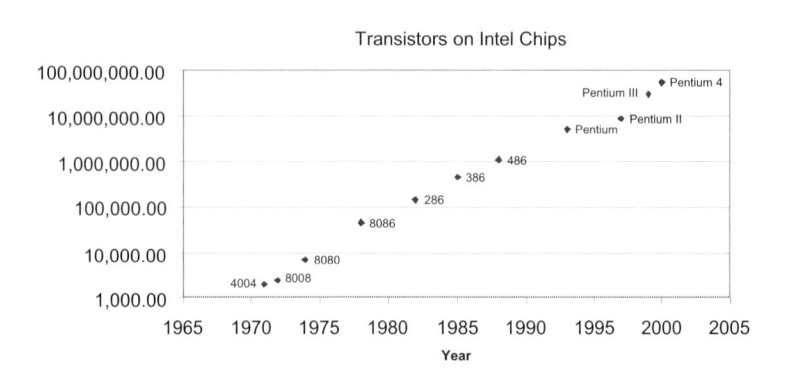

4.1 The continuing growth of CPU power over 30 years based on Intel computer chips

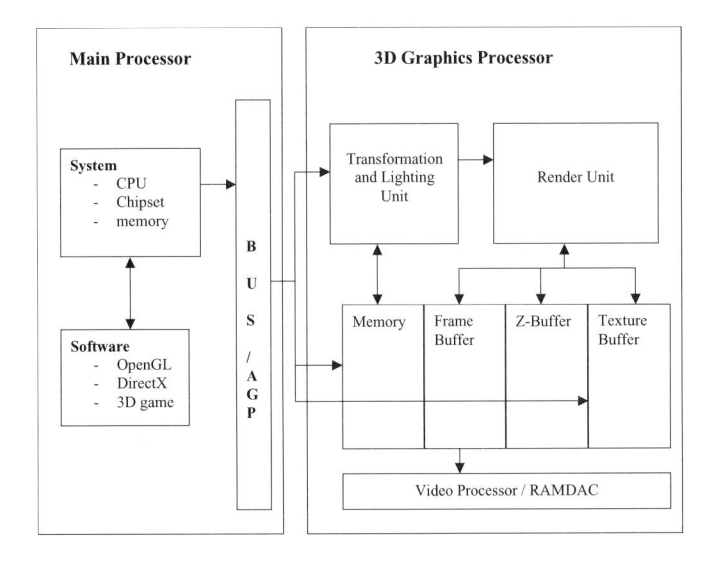

4.2 The basic configuration of a PC with graphics accelerator card

(now SGI) of their Reality Engine, the chase has been on for faster graphics performance through specialized graphics processors. This is typically an extra processing card added into the computer and linked to the CPU (on the mother board) via either the standard internal communications bus (e.g. PCI) or a specialized graphics port: e.g. Advanced Graphics Port (AGP).

On the left of Figure 4.2 is the motherboard with the chipset, the memory, the CPU and AGP slot. In terms of graphics performance this can be seen as one entity, sometimes called the platform. Here, the software reads the dataset describing the virtual environment. A viewpoint is specified through the user interface and the platform determines all the objects, lights, textures, behavioural rules and so forth that are relevant to the creation of the next image.

The vertices which describe the basic geometry of the model are passed to the transformation and lighting unit. Here they are transformed from world coordinates to the perspective coordinates of the camera/viewer. Basic lighting calculations can be done at the same time. The results of this process are passed to the render unit. The 3D-card then completes the rendering and gets the resulting picture from the frame buffer to the screen. Ideally, the main processor and the graphics card can do their respective jobs at a similar rate. Whichever is the slower will determine the frame rate of the computer. Depending on the application, the graphics card often has much more work to do and so is the key constraint.

The frame rate of the graphics card is also dependent on screen resolution. One of the key specifications of the graphics card is its fill rate. This describes the number of pixels that a card can render in a given amount of time. The 3D-chip has to 'render' each pixel of a frame before the frame can be displayed. Nowadays, 3D-chips have several rendering pipelines that can operate in parallel. Such a pipeline is usually able to render one pixel per clock cycle. Thus, the maximal pixel fill rate is the 3D-chip clock speed times the number of rendering pipelines times the number of chips. If this adds up to, for example, one billion pixels/second then one can quickly determine a frame rate based on screen resolution. Two million screen pixels can, in theory, be refreshed in the frame buffer 500 times per second. However, we do not actually see speeds like this because the platform cannot delivery data for the frames that fast and there may also be a limitation caused by the speed at which data can be moved into or out of memory (the memory bandwidth).

In Figure 4.2 the load on local memory becomes clear. There is a great deal of passing of information between the renderer and the memory.

1. The local memory hosts the *frame buffer*, which must store two or three pictures at once (the one on the screen, the one waiting to go on the screen next, and sometimes a third). If 32 bit colour is being used, that means up to 12 bytes/pixel. The frame buffer needs to be accessed by the rendering unit for each pixel several times.
2. The *Z-buffer* (for determining which objects are visible and which are hidden) is also as big as the screen resolution times the Z-buffer depth. It, too, is accessed several times per frame depending on the number of objects being rendered.
3. Then there is the *texture buffer*, which holds compressed or uncompressed textures for fast access by the rendering unit. Textures may need to be read several times for each pixel.
4. The transformation and lighting unit requires memory access but this is a lot less than the demands above.
5. Finally, there is the RAMDAC (Random Access Memory Digital-to-Analog Converter), which needs to read the front frame buffer to display it on the screen. The higher the resolution and the higher the display refresh rate, the more often the RAMDAC has to access the frame buffer.

Consequently, if you compare graphic card specifications, you commonly see a lot of emphasis being placed on the memory clock speed and, especially, the memory bandwidth of each card.

As an example, one current graphics card has a 17.6 gigabyte/second memory bandwidth. However, it can output at resolutions up to 3048 × 1024 – over 3 million pixels. So, if the frame and Z-buffers require 24 bytes/pixel and are accessed on average 3 times per pixel draw then the potential frame rate is already limited to around 80 fps without taking account of the texture buffer or the transfer to screen.

In terms of graphics performance we can look for some equivalent of Moore's law for graphics processors. If we simply take pixel fill rate (possibly not the best indicator but one that has been quoted for some

Table 4.1 The changing specifications of computer graphics

	PC cards			SGI solutions		
Year	Fill rate (pixels/sec)	Draw rate (polys/sec)	Card	Fill rate (pixels/sec)	Draw rate (polys/sec)	Engine
1992				320 M	210 K	Reality Engine
1996	26 M		3D Rage II			
1997				400 M	11 M	Infinite Reality
1998	250 M		Riva TNT			
1999	400 M	15 M	Geforce 256			
2000	1.6 G	25 M	GeForce 2			
2003	2.6 G	275 M	Radeon 9700	3.8 G	141 M	Infinite Performance

years) we can see something of the trend (Table 4.1). The bottom line is that today's PC cards costing a few hundred dollars exceed the fill rates of all but SGI's million-dollar RealityMonster of earlier years. It appears from this table that there has been a tenfold increase in performance over the past five or six years. That is, a doubling rate well under 12 months.

EFFICIENT MODELLING AND RENDERING
OF LANDSCAPES

*Oliver Deussen, Carsten Colditz, Liviu Coconu and
Hans-Christian Hege*

Effective generation of virtual landscape in real time depends not only on fast hardware but also on efficient algorithms for drawing the land and the vegetation. In botany, such models can be used for simulating plant growth and genetic expression, for evaluating mathematical models applied to backscattering measurements and for visualizing spatio-temporal processes. In the movie industry they are used for modelling special effects or special plants that cannot be found in nature. In computer games they are used for synthetic backgrounds. The different applications need different models but, generally, the desire for very detailed visual plant descriptions grows as fast as computers are able to handle those complex models.

Development of plant models

In recent years, several important advances have been made to create such realistic plants models efficiently. Two major mechanisms are described in the literature, on the one hand procedural methods are used, parameterized algorithms that generate plant geometry for one plant or a small set of plants. Various algorithms were proposed (Aono and Kunii 1984; Oppenheimer 1986; Bloomenthal 1985; Holton 1994), but few are still widely used.

One the other hand, rule-based systems describe the plant geometry by a set of rules that are applied to create a complex model from a simple initial state. The most prominent approach is known as L-Systems, developed by Prusinkiewicz and Lindenmayer (1990). Here, a textual rule basis is used for describing the plant. The rules do string rewriting in a given text. Starting from an initial word, the sequence grows until a given number of rewritings is performed. In a second step, the final string is interpreted graphically to produce the geometry. Over the years these systems were extended by mechanisms for interaction of plants with their environment (Mech and Prusinkiewicz 1996) and by introducing positional information (Prusinkiewicz *et al.* 2001). The latter extension allows the user to edit the models quite efficiently. Spline functions can be used to vary important parameters of plants along their growth axes. Instead of programming L-System rules manually, the user is able to change the shape of plants using parameters based on these functions.

Visualization technology

4.3 Steps in creating a sunflower: real leaf textures are scanned and projected on the surface of a leaf element represented by a leaf component. Arrangement of plant parts is performed by multiplying components: as shown in the blossom. Putting all together, a set of ten components is able to represent the whole plant

A similar editing technique was proposed by Lintermann and Deussen (1999) in combination with the plant modelling tool 'xfrog' (www.greenworks.de). Here, a mixture of procedural and rule-based methods is used to model a plant: procedures compute the geometry of plant parts and are combined by a simple rule-based mechanism. The procedures are represented by components. The user connects a set of components to describe the structure of the plant. The algorithms are controlled by graphical user interfaces on the basis of spline functions. A sample plant is shown in Figure 4.3.

Modelling synthetic landscapes

Terrain, human artefacts and a set of plants form a landscape. For modelling plant societies, growth patterns and statistic values of the plant populations are obtained from nature and are represented by the computer. The main problem is simply the number of plants and the huge amount of geometry which is necessary to model vegetation. A single square metre of a meadow contains thousands of plants and needs millions of triangles to represent it realistically. Millions of plants must be combined to represent a square mile. Deussen *et al.* (1998) describe an open system that is able to model and render complex plant scenes with hundreds of millions of triangles (Figure 4.4). In a recent

Understanding visualization

4.4 A synthetic landscape modelled by Bernd Lintermann using xfrog and the open system described in Deussen *et al.* (1998) (image reproduced in the colour plate section)

test, the ray-tracing algorithms rendered the image in about one hour on a sixteen processor SGI computer – far away from interactive rates.

In Deussen *et al.* (2002) the rendering performance for the same scenes was improved to several frames per second – now using graphics hardware and a specially designed level-of-detail (LoD) algorithm. The plant is represented by geometry, if the virtual viewer is close, and by a set of points and lines if far away. Points are used for compact objects such as leaves, lines for long and thin objects such as branches. Both sets are obtained by randomly sampling the plant's surface. For a given distance to the plant or a given size of the projected plant model, respectively, a specific number of points and lines is displayed. This number decreases with distance such that the number of points and lines always represents the plant model faithfully.

The problem with most of these LoD schemes is the computing time necessary for each frame. In our case we solved the problem by randomly reordering the point and line sets for each plant and storing them in graphics memory. This reordering is done in such a way that each leading portion of a set always results in a valid approximation of the plant geometry. The amount of points and lines shown for each plant is now computed by a simple formula and results in a number which tells the graphics card how many of the points and lines for a plant model have to be displayed. Doing so, no transfer of data to the graphics card is necessary during rendering, all the approximation data is stored on the card in a pre-processing step.

Figure 4.5 shows some examples of plant approximations. In Figure 4.5(a)–(e), a plant model is represented by geometry and by several sets

Visualization technology

4.5 Level-of-detail description of a plants: (a) geometric description of a pine tree; (b) representation by 13 000 points; (c) 6500 points; (d) 3250 points; (e) 1600 points; (f) several plants, the models in the background are displayed by point sets; (g) line approximation of a plant with thin leaves

of points. In Figure 4.5(f), a set of plants is shown, the plants in the background are represented by points. In Figure 4.5(g), a line approximation of a plant with thin leaves is given.

Showing larger scenes requires some additional effort. In a plant population of several million plants it is very time consuming to represent every plant even if only one point is shown. This arises from the fact that each model has to be visited and the model representation has to be read. Much cheaper (in processing time) is to represent a number of plants – a square metre or larger – by a new virtual plant description. This is done by building a spatial data structure for larger terrains that stores squares of the terrain in the form of virtual plants. Doing so, even larger scenes can be displayed interactively.

On the other hand, sometimes a very large plant has to be subdivided into several virtual plants. If the camera is close to such a plant, the near leaves have to be represented as geometry while the far leaves still can be approximated. As our LoD method works for the plant as a whole, it has to be split into parts to improve performance.

Currently, we are working on hierarchical spatial data structures to enable interactive rendering speed for arbitrary landscapes and – in the more distant future – for a whole synthetic planet.

Rendering terrain

Another time consuming task for the computer is to represent terrain. Usually terrain data is represented by irregular sets of triangles (TIN) or regular grids for the surface and large textures for its visual appearance. Geometry and texture can be of enormous size. Often it is not possible to store the whole data in main memory and therefore efficient hierarchical caching algorithms have to be used. More than that, in many cases it is necessary to model important parts of the scene with a higher complexity than others. This can be represented easily by TINs but requires some effort using regular grids. In this case, a hierarchical representation is used: a basic grid represents the scene, but important parts do have their own local grids that replace the basic grid.

For display, another kind of hierarchy is required: the scene is divided into parts using a so-called quadtree. In the first step the complete scene is divided into four parts of equal size. If the geometric complexity of the parts is above a given threshold, they are further divided into four parts. All parts are stored as knots in the quadtree. Doing so, it is necessary to cut the geometry of each part at the border. Again, this is easy for TINs but a harder job if a grid hierarchy is used. In this case, all the local refinements must be clipped also.

Now geometry is represented hierarchically for each part of the quadtree. This is done by representing the highest order knots by a base mesh which is refined by the data associated with the child knots. The algorithm ensures that the error in the altitude is always below a threshold for each approximation inside the quadtree. If the virtual viewer moves towards a point in the scene, closer and closer approximations are computed by moving down the tree hierarchy. Moving horizontally requires reading of new parts of the base meshes and refining them. Both operations can be realized using solely local refinements of the geometry and by ensuring a minimal data stream. Our approach is based on a work presented by Thatcher (2002), which allows complex terrain data to be handled interactively.

Conclusion

It is still a difficult task to represent complex virtual landscapes interactively. The developed algorithms for modelling and rendering of single plants seem to work well for the quality needed in outdoor scenes. However, much more work has to be done to handle all the data involved in this process.

In this context, the rendering capacity of modern graphics cards is not so much the problem, much more important is the bandwidth between hard disk, main memory and graphics memory to transport the enormous amount of data needed to represent outdoor scenes. Theoretically, buses like AGP 8X can transfer up to 2 gigabytes per second for uploading onto the graphics board but in practice values are much lower. Therefore, efficient data representation and compression techniques have to be developed to reduce the amount of data to be transferred. More and more operations can be performed in the graphics card which will help to solve these problems.

Remaining problems involve animation and interactive editing. So far our data structures are quite static and do not work well with dynamically changing data. We are able to cheat in order to simulate wind, but animations such as growth of plants is not possible yet. Also, the user has limited possibilities to change landscapes interactively. This is because the geometric description of many plants is combined to an overall representation of the scene to enable interactive rendering. Single plants can be added here, but after changing substantial parts of the scene, pre-processing must be performed again before display.

In the future we will work in these directions. Our goal is to develop a front-end processor for GIS that allows the user to visualize complex landscapes with rich vegetation and all other artifacts needed for landscaping.

USING GAMES SOFTWARE FOR INTERACTIVE LANDSCAPE VISUALIZATION

Adrian Herwig, Einar Kretzler and Philip Paar

Introduction

The rapid development of hardware and software for interactive visualization is triggered by the fast-growing market for computer games and underwritten by millions of enthusiastic computer game players. Many computer games can simulate virtual environments, e.g. synthetic landscapes, close-to-reality in real time on personal computers or game consoles. This rapid development in computer game technology is almost unnoticed by the users of professional CAD, GIS and illustration software. While many of these games are objectionable because they glorify mindless violence, landscape planners should have an eye on these developments because some components of the software may be useful for their purposes. Many of today's popular games include entire outdoor environments. This technology can be applied in landscape planning and landscape architecture.

Games software and technology

As the first computer systems with graphic features became available, programmers tried to visualize synthetic environments in real time. Therefore, game engines, computer programs that contain the core algorithms controlling a game and handle the display of a game's environment, have been developed. One key feature is the ability of the user to interact with the virtual world. An input device is used to navigate within a scene or to control the course of a game's story. The development and adoption of the mouse, which could operate a switch or aim a target was essential to games development.

In 1993, when the first computer games using graphics and mouse input reached the market, another innovation was made. Some games offered an interface so that users could create their own *levels* – custom-made three-dimensional environments. Through the years the handling of these CAD-like *level editors* has been much improved.

Lots of game engines have been developed since then (Herwig and Paar 2002). They all have specific features corresponding to the intention of the particular game. Some years ago game engines were rather specialized in the kind of virtual environment they supported. Engines for flight simulators could show large terrains but did not require an accurate display of the scenery. Indoor three-dimensional games or

Visualization technology

4.6 Virtual measures, screenshots in (top right) first person view; (above) third person view with an avatar; (left) top view; (top left) interactive design of a tree row

top-down games, with a downward angle point of view, made little or no use of geometry reduction and level of detail (LoD) management. Engines for unconstrained ground games let you look up or down, and get close to the surface. They must allow a wide range of distances and LoDs. Some have third person display modes (Figure 4.6), but allow the full navigational freedom of a first-person view. The player is represented by a virtual figure, an 'avatar'. The avatar's movements are restricted by physics, e.g. terrain following and collision detection.

Today's games software combines several engine modules, each with a specific functionality. A terrain engine creates the ground, while a physics engine controls the motion behaviour of characters and objects including realistic simulation of gravitational force and kinetics. There are increasing interaction and movement capabilities, which let avatars move, take,

Understanding visualization

leave or change objects within the virtual environments, or which allow avatars to climb, crouch, swim, dive or fly. Artificial intelligence (AI) allows intelligent behaviour of characters or other game entities that are not directly controlled by a person playing the game. Many game engines offer Intranet- and Internet-server-based multi-player support including chat features. Only a few have extensions for database access.

Computer hardware, especially the graphics processing unit (GPU), have improved so much (see 'The significance of computer power', p. 51) that features like dynamic lighting and deformation of objects can now be generated in real time. Games technology is optimized to support emerging features of the GPUs. This increases the possibilities of interaction and makes games more versatile. In addition, recent advancements in programmable GPUs make cinematic special effects available to games software, e.g. facial expressions of characters, realistic explosions and weather effects. Many formerly software-based processes are now per-

4.7 Representation of (top) water with pixel shader; (right) water curtain on a plexiglass wall (images reproduced in the colour plate section)

formed by the GPU, like vertex- and pixelshaders for realistic outdoor surface lighting. Pixelshaders enable close-to-reality water simulations, e.g. dropping a stone in the water causes wave circles (Figure 4.7). Texture animations can simulate the water flow using transparent materials (Figure 4.7). Three-dimensional sound makes the virtual environment more authentic, especially with a surround audio system.

Games software for the landscape planning and design process

There have been many examples of professional game-technology-based visual simulations, e.g. in military www.americasarmy.com), architecture (www.arct.cam.ac.uk/research/pubs/pdfs/rich00a.pdf), heritage (digitalo.com/deleon/vrglades/#) and archaeology (conitec.net/gallery.htm) applications. However, not every successful game engine with rich features is suitable for landscape visualization. The game software needs to support first and third person views, bird's eye views and large terrains. These enable walk-throughs and plan views with near-immediate visual feedback to design and conceptual suggestions. The developer of an application can program interactive options, enabling the user to change the compositional characteristics of a scene. Materials can be tested immediately through changes in texture. Height parameters for objects or terrain can be adjusted interactively or the objects can be switched on and off (Figure 4.8). Using three-dimensional game engines for landscape and garden visualization requires a fundamental knowledge of three-dimensional designing. The engines provide little support for working with different levels or for import of objects from other software. At this stage, using game engines for landscape visualization requires special strategies for dealing with these deficiencies. For example, a working strategy in level design is necessary and trial-and-error methods and external converters

4.8 Interactive modifier for the fountain height (images reproduced in the colour plate section)

are often necessary to import objects from CAD, GIS or three-dimensional modelling software. On the other hand, three-dimensional game engines are versatile and low-cost tools, and a worldwide user community provides help and resources via the Internet. Some companies are now providing enhanced editing tools through 'professional' versions.

Using these editors, planners (developers) can publish and distribute their landscape visualizations freely without having to pay any royalties or licence fees. An interactive presentation can be distributed via download, CD-Rom or DVD. The planner could either make an executable program available to the stakeholder, customer or citizen or use the visual simulations in civic participation processes and for presentation systems (Herwig and Paar 2002). Games software-based applications can also be used for education or to promote ecological awareness. Orland *et al.* (1997) proposed an 'interactive game-like visualization', which engages users in understanding how the ecosystem is working and enables interactively changing environmental parameters. They expand the idea of 'playing the environment', as exemplified by computer educational programs such as SimCity and SimFarm, but with realistic and manipulative data and models. Considering increasing bandwidth, applications for online planning or for citizen participation via Internet are becoming more attractive.

Creating a virtual environment

The following section describes the procedure to create a real-time application using a commercial software package. At first, the user has to prepare the existing base data, which is available as GIS and CAD data. For example, a digital elevation model (DEM) could be on hand as a distribution of elevation points, as a triangle mesh or as contour lines. These terrain data have to be converted into a 256-grayscale bitmap graphic, very similar to the USGS DEM format, where darkest means lowest and brightest means highest. Then the model editor is used to import the bitmap file and create a terrain model. A GIS shape, representing the land use or an aerial image can be used to texture the surface of the terrain. Therefore, the data must be converted into a bitmap graphic.

Three-dimensional objects like trees and animated models can either be created with the model editor, or imported from external three-dimensional modellers. If static geometries ('brushes') like buildings and bridges are only available as two-dimensional CAD drawings or GIS shapes, they have to be modelled three-dimensionally in an external three-dimensional modeller and converted to an importable format. During preprocessing, visibility, lighting and collision detection calculations are applied to these brushes (models). The model output formats may also contain information about a model's movement abilities. In the level editor everything is assembled: terrain and three-dimensional objects, brushes and the environmental properties like lighting, sky and weather effects. The finished level is then preprocessed, in other words the files that contain all scene information run through a compiling procedure. This is how the game is made suitable for real-time, interactive play.

First, the level geometry is split off, i.e. every surface is divided into squares and triangles and the empty space is divided into so-called *leafs*. A *binary tree* is built to arrange all the elements. Collision conditions for the camera (and the avatar) are determined. Next, a light map and a shadow map are generated. These contain all implemented lighting information. Finally, a *potential visibility set* (PVS) is computed to help the game engine to decide what element to draw or to skip in the view. Complex three-dimensional objects can be set with different LoDs (see 'Modelling synthetic landscapes', p. 57). Compiled levels can then be explored in the game engine's stand-alone three-dimensional player in real time.

Conclusion

Three-dimensional game development and landscape planning share an interdisciplinary approach and the need for visualization. It is questionable whether non-interactive animations are worthwhile in the landscape planning and design process. Using real-time games technology, there is no time-consuming image processing because rendering is hardware-accelerated by the GPU.

Games software-based visualizations provide persuasive means of advertising, and they might be used to optimize planning and design. Three-dimensional games software provides interesting and low-budget alternatives to professional three-dimensional landscape visualization software. They are optimized for real-time navigation in virtual scenery. However, professional features like a GIS-data interface or georeferencing are still missing, and realistic representation of the vegetation is lacking. Features like artificial intelligence and multi-user support are resources, as yet barely touched, to create highly sophisticated interactive landscape visualizations.

PRESENTATION STYLE AND TECHNOLOGY

Ian D. Bishop and Eckart Lange

The simplest and most common form of visual display is a single rendered image on a general-purpose computer monitor. Sophistication can increase in terms of both the style of the presentation and presentation environment/technology. As discussed above, the rendered world may be presented as either static or interactive. There are also in-between options like panoramas and animations.

The display environment can range from a printed page to multiple screens to head-mounted devices. In addition, a local machine could be storing the data and doing the processing while the visualization is delivered via a network such as the Internet. These options are discussed in the context of human perception under the assumption that accurate perception underpins good decision making.

Panoramas

The QTVR format allows a full 360° panorama to be viewed on a normal computer monitor. The user can control their view angle and zoom extent and undertake quite a realistic exploration of an existing location or one generated by computer (Figure 4.9). Bishop and Hulse (1994) used 360° video panoramas in visual quality assessment, arguing that to effectively assess a location rather than a specific view people need to be able to see the whole environment. Comparing on-site views at a panoramic location with normal 4:3 ratio photographs, Palmer and

4.9 Computer generated panorama of the Seewis area, Switzerland (from Lange *et al.* 2001)

4.10 Route of both the virtual and real walk used by Bishop and Rohrmann (2003)

Hoffman (2001) found that respondents 'were not able to objectively ignore the dramatic context of the panorama when they were directed to evaluate that portion of the on-site view represented in the photographs'. Meitner (2004) also pursued this argument but found that it may not be necessary to show a linked panorama because four photographs shown together and covering the whole view can provide very similar results.

Animations

Animation can involve either movement of objects within a scene, movement of the camera or both. In presentation of projects for public comment it is a common option to provide a fly-over or walk-through of a site to give a more complete picture of the environment being considered. Bishop and Rohrmann (2003) have undertaken a series of experiments designed to determine whether such animations are an acceptable surrogate for taking a walk through a real site (Figure 4.10). Their results were not definitive and suggested that sensual realism remains an elusive goal (as also indicated for still renderings by Appleton and Lovett (2003)).

One of the most effective uses of animation in landscape research was the work of Hetherington *et al.* (1993) who animated stream flow using an analog video overlay technique in order to determine how different flow rates in wild and scenic rivers affected scenic quality estimates.

Interactive visualizations

With the power to redraw the view many times each second came the concept of Virtual Reality (see also Chapter 2). Researchers have begun to use VR technology to create virtual environments or landscapes as experimental platforms for measurement of environmental response – especially navigation (Bowman *et al.* 1999; Darken and Sibert 1996; van Veen *et al.* 1998). In the landscape context, Bishop and others (Bishop 2001; Bishop *et al.* 2001a; Bishop *et al.* 2001b) have argued that an experiential (Zube *et al.* 1982) or phenomenological (Daniel and Vining 1983) approach to landscape perception research is now possible and holds more promise than the classical psychophysical approach (Daniel and Vining 1983) using verbal responses.

Such work need not be confined to specialized computer systems since the average desk-top computer can now access interactive three-dimensional worlds over the Internet. Using protocols such as VRML (Figure 4.12(b)) and Adobe Atmosphere (Figure 4.13) virtual worlds complete with avatars (representations of 'residents' or fellow visitors in the space), and interactive elements can be widely accessed (Champion 2003; Nadeau 1999).

The concept of presence is a very important one in virtual environment research. For such worlds to be effective, a level of presence (and the related concept engagement) must be established. An extensive review of these concepts can be found in Draper *et al.* (1998), who identify three types of presence in the literature: simple, cybernetic and experiential. According to Slater and Steed (2000), the first is simply the ability to operate in the virtual environment, while cybernetic presence is determined by aspects of the human–computer interface. Slater and Steed (2000) concentrate on the third approach, experiential presence, defined as 'a mental state in which a user feels physically present within the computer-mediated environment'. They found that action through appropriate whole-body movements, is strongly associated with a higher sense of presence. This suggests that for a full experiential experience of landscape, and effective use of the virtual environment for landscape research, control of movement through the virtual environment using body movement will generate more presence and thus more experiential legitimacy than use of a mouse, joystick or other control device. Similarly, Draper (1998) writes that '... immersion is the degree to which sensory input to all modalities is controlled by the SE (synthetic environment) interface'.

Multi-sensory presentation

The concentration on visual representation in presenting landscapes for public assessment is justified by the dominance of vision over other senses (see Chapter 1, 'Why visualize?', p. 3). Nevertheless, some researchers have also added different sound conditions to the visual stimuli in landscape perception research (Anderson *et al.* 1983; Carles *et al.* 1999; Esposito and Orland 1984; Hetherington *et al.* 1993).

There are also environmental planning circumstances in which senses other than vision may be of greater importance – consider the sound of a waterfall, the smell of a pine forest, the feel of sea-spray or the hum of traffic. Ideally, we will extend our virtual worlds to provide appropriate stimuli to all the senses. Langendorf (2001) summarizes recent developments in auditory, tactile and olfactory simulation.

Printing

Printed versions of still images may be used in a number of different circumstances. Hard copy is the easiest solution if the images are to be used in the field for a comparison of current conditions with the simulated images (Bishop *et al.* 2003). Larger format printing can also be used in a public display environment and the public then invited to complete questionnaires relating to the images, on a drop-in basis (Petschek 2003). Prints remain the only effective presentation medium when a Q-sort approach to landscape assessment is being used (Lange 2001).

Single screen

Monitors come in a variety of sizes and technologies. The smallest sizes are found in specialized devices like mobile phones. The most common monitors are those normally used by an individual in conjunction with a desk-top or lap-top computer. The largest are for use by a small audience and include video walls and plasma panels.

A single monitor or data projector can show more than just still images. Animations and interactive media are also options. For the single user, the computer display is sufficient – and may be more appropriate for response that requires interaction (Bishop *et al.* 2001b) – but for a larger audience a projected display can offer high graphic quality even in daylight.

The individual monitor is commonly distinguished by size and resolution and given an acronym on the basis of resolution (see Table 4.2). All these, except HDTV, have an aspect ratio of 4 horizontal to 3 vertical. Unfortunately, this is nothing like our normal vision which has a much larger horizontal to vertical ratio. Unless enlarged to a very great extent using specialized projection systems (e.g. IMAX) this aspect ratio only fills a small portion of our field of view. HDTV with its 16:9 ratio is closer to filling our field of view but is still not sufficient for 'immersion' in the display.

Understanding visualization

Table 4.2 Typical monitor resolutions

Name	Resolution
VGA	640 × 480
SVGA	800 × 600
XGA	1024 × 768
SXGA	1280 × 1024
UXGA	1600 × 1200
HDTV	1920 × 1080
QXGA	2048 × 1536

Multi-screen immersion

To achieve a sufficient field of view to give people a feeling of being in an environment, we need multiple displays. These may be two monitors, one for each eye, as in a head-mounted display, or multiple projectors filling a field of view on a special screen (Figure 4.11). The screen may be a single curved unit allowing a panoramic view of the projected environment, or it may be a set of flat screens. If the flat screens are orthogonal and surround the viewer then this is the CAVE environment (Cruz-Neira *et al.* 1992). A CAVE will usually have from four (three walls and the ceiling) to six sides operational depending on application and budget.

4.11 Options for immersive display: (below) head-mounted display – this one is bulkier than most but also provides a wider field of view and higher image resolution; (right) the panoramic screen – in this case three projectors and three flat screens (note the shadow problem that comes from front projection)

In environmental visualization there has been a strong tendency to work with panoramic displays rather than using a CAVE or HMD to provide a sense of immersion. The reason for this trend appears to be a mixture of preference for multi-user access, ability to work collaboratively while viewing the display, lower nausea risk and also cost. A number of landscape research laboratories around the world now have, or have access to, multiple projector display environments. The hardware is typically a curved or three-part screen, three or more projectors and then either multiple linked and synchronized computers or a single computer with multiple output channels. A recently introduced product offers a flexible environment in which the side panels can be folded in to form a partial CAVE or left extended for panoramic projection (see Chapter 1, Figure 1.10(b)).

The cost of such an environment has dropped dramatically in recent years. Many labs were initially based on specialized graphics workstations (especially those made by Silicon Graphics). The option for some time was to link lower cost computers via software. Recently, however, multiple output graphic cards for PC-level machines have appeared. High performance output to three projectors is now available for roughly one-five-hundredth of what might have been paid just six years ago (see also Figure 5.7, p. 96, for a typical screen set up and the section 'Helping rural communities envision their future', p. 145, for an example application).

Stereo display

Stereoscopic display is achieved by giving each eye a different image. The computer renders the scene from slightly different viewpoints. By putting these viewpoints further apart than the separation between our eyes, the stereo effect can be exaggerated.

For each eye to see a separate image it is necessary to use either two separate screens (as in a head-mounted display) or to provide the two images at different times on a single screen (whether a monitor or through a projector) and differentially control the access of the view to each eye. As all these techniques require two distinct images to be rendered, this halves the effective refresh rate of the system.

So-called *passive stereo* may be achieved by wearing lightweight glasses which filter out part of the signal on the basis of colour or polarity. The colour – anaglyph – approach is less effective because the colour of the rendering is necessarily distorted. In the polarization approach, one lens accepts only horizontally polarized light while the other accepts only vertically polarized light. The two images are projected onto the same screen by two LCD projectors which each produce signals of single polarity.

The alternative is an *active stereo* approach in which the images for the left and right eye alternate on the screen and shutter glasses are syn-

Understanding visualization

chronized with the screen display such that the left eye sees the left images for 1/60 of a second followed by the exposure of the right image to the right eye.

Some firms are experimenting with autostereo displays which work without the assistance of glasses. A special lens is installed in front of an LCD screen to restrict visual access for parts of the screen for each eye. The left and right images are vertically interlaced with each other, which effectively halves the screen resolution.

World Wide Web – making three-dimensional visualization widely available

The growth of the Internet has been charted by many authors. The major trend early in the twenty-first century is for increased domestic broadband access. While interaction with complex models was a slow and frustrating business with a regular telephone line modem, broad-

4.12 Examples of the way in which the Internet can a provide a mechanism for exploration of real or simulated environments:
(right) example of a page containing a QTVR image in which a 360° panorama can be explored from a single viewpoint;
(bottom right) complex virtual models can be represented in VRML and explored from any position or angle

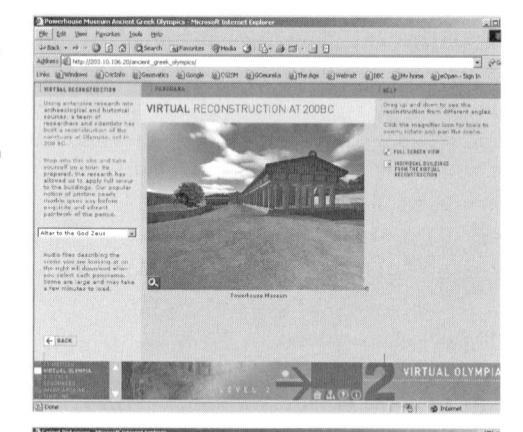

4.13 The combination of a three-dimensional model, an avatar representing the user and another who is either a additional visitor to the space or an agent acting as a guide; also a place for chatting to the computer or to other visitors

band opens up many possibilities for widespread access to viewing and interaction with complex image formats or models (Figure 4.12).

New protocols are emerging which go beyond VRML for interactive potential. It will be increasingly possible for people to meet other people (or at least their avatars) in virtual worlds. In this way exploration, negotiation and decision making are possible in a shared context by people at wide distances. Figure 4.13 shows a partial reconstruction of the Mayan city of Palenque. In the view is the avatar of a visitor to the site. Also on screen is a place for chatting (text-based) with the other visitor. A variety of other controls and features exist to provide a rich virtual environment.

The Internet potentially provides a much wider audience for planning-related opinion studies than face-to-face surveys. Data collection is automatic without the need for later entry. On the other hand, less control is possible over the display quality of survey conditions while those responding typically do not form a representative community cross-section. These issues are comprehensively reviewed by Wherrett (1999).

There are various ways in which this response recording can be organized. The simplest approach is for the web page to be designated as a 'form', then there must be a program or script of some kind that is activated when the form is submitted and this processes the data in some way. For some time the dominant procedure was for the survey designer to create a C or PERL program which wrote the responses to a file. More recently the opportunity to have each response e-mailed to a designated address has been widely used. The introduction of PHP (a self-referring acronym for PHP Hypertext Prepocessor) has made life easier for non-programmers because the control script can be embedded in the HTML script which defines the web page.

In some cases a web site can provide a kind of informal survey of popular will. Dodge (1999) reviewed the way in which the cyber environment called AlphaWorld has developed as people were free to choose a site at which to build a virtual home.

Animations can also be delivered over the web in various formats. The simplest is the animated GIF file. A series of images can be made into a single GIF image (using low cost software) which can be made to loop through the images at any specified speed. This is ideal for any sequence which repeats after a reasonably small number of frames. The GIF file can then be installed on a web page. Bishop (2002) used this approach to assess the perceived size of a wind turbine relative to a stationary tower. Longer, non-looping, animations can also be played directly on the web page (with appropriate software plug-ins) but may take a long time to download over conventional telephone lines. This may be a severe disincentive for potential survey participants.

Understanding visualization

Human factors

Much of the landscape visualization development over the past three decades has focused on faster hardware and better graphics and, hence, a greater degree of realism. In the past five years, on the other hand, it has been widely recognized that not only do we need to be able to simulate realistic looking landscapes (although the ultimate level of realism required is still open to question) but also that we have to make our simulations smarter. They need to link to other technologies and provide users with great flexibility to become widely accepted and used. Paar and Buhmann (pers.comm.) recently surveyed over 300 users in both private and public sectors in Germany about the features most desired in three-dimensional landscape visualization software. Paar and Buhmann found (Figure 4.14) that ease of learning and interoperability were the most highly regarded features, while factors such as speed, interactivity and photorealism rated quite low.

In recognition of the importance of human factors to the success of visualization tools, MacEachren *et al.* (1999) state:

> A working hypothesis behind much of the research in visualization in scientific computing (ViSC) over the past decade is that the most successful visual representation methods will be ones

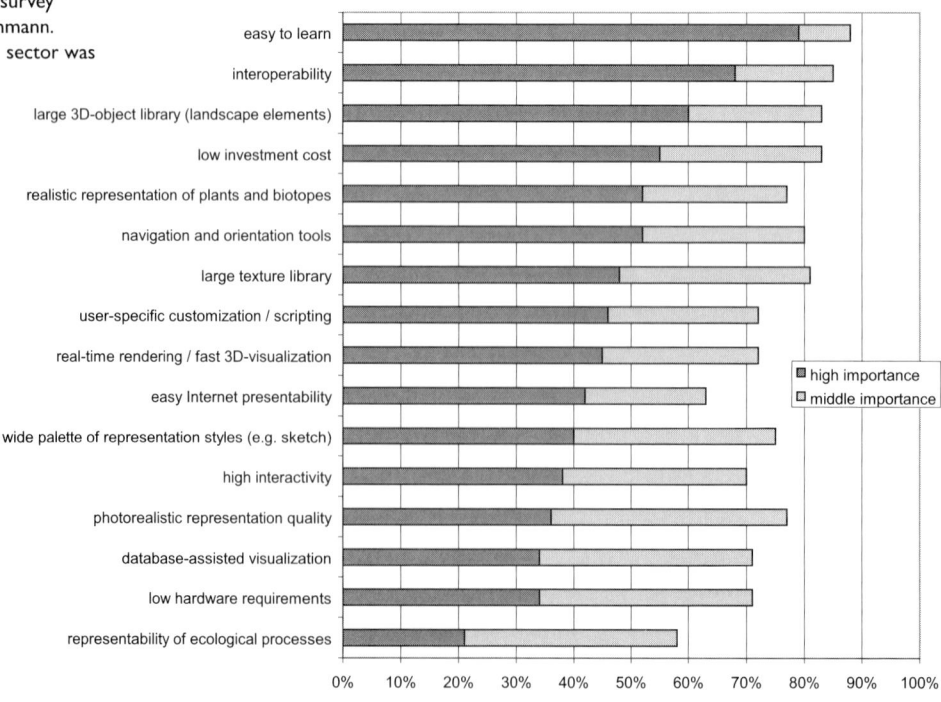

4.14 Demand for features of three-dimensional landscape planning software based on the responses of the private sector (landscape architecture firms and engineering offices) to the survey conducted by Paar and Buhmann. The response of the public sector was very similar

that take the fullest advantage of human sensory and cognitive systems developed for interacting with the real world. As a result, emphasis in ViSC has been on 3D dynamic displays and realism applied to the representation of objects, particularly objects that have visible form in the real world (e.g. the human body, aircraft wings, thunderstorms). Extension of these methods for use with VE technology requires only modest changes conceptually (although there are technical challenges).

CHAPTER 5

VALIDITY, RELIABILITY AND ETHICS IN VISUALIZATION

Stephen R.J. Sheppard

The crystal ball

Computer visualization of landscapes in four dimensions constitutes a 'crystal ball' capable of showing us views into the future. The quickening pace of technology, driven primarily by the entertainment sector, promises continuous 'improvement' in visualization capabilities: a faster, more realistic, more sophisticated crystal ball. However, do we *need* a better crystal ball? In the context of planning practice, do the inevitable 'improvements', responding to available technology and popular demand, actually make visualizations better?

Addressing this issue requires us to ask a series of questions:

1   What does 'better' mean?
2   How good are visualizations already?
3   If current visualization technology is not good enough, how should it be improved?

This chapter is based on the premise that emerging visualization systems should respond to tangible social and professional needs, not merely to commercial opportunities and popular expectation; and consequently, as the underpinning of those systems, we need to seriously consider issues of validity, reliability and ethics in the design and use of these potentially powerful tools. As Palmer and Hoffman (2001, 149–161) put it, in the context of the field of landscape perception and assessment: 'to be good, our work must be both reliable and valid'. This chapter, therefore, focuses on these issues of quality, rather than on other important issues of the utility of landscape visualization technology (Table 5.1), such as cost-effectiveness or operational 'functionalities', which are likely to be taken care of by market factors sooner or later.

Understanding visualization

Table 5.1 A simple framework for evaluating aspects of landscape visualization

		Components of visualization system	
		Tools	Support infrastructure
Visualization issues	Quality	*Are the focus of this chapter*	
	Utility		

Various potentially desirable outcomes of improved visualization systems can be postulated, going beyond benefits to the preparer such as greater efficiency and usability, to include improved communication of information and support for better decision making. This chapter reviews conceptual principles for 'good' visualization systems to aid in evaluating visualization quality, drawing from available research findings, empirical evidence from practice, and perceptual theory, before providing suggestions for design of new visualization systems to achieve the desirable outcomes identified.

The need to establish a better support infrastructure to guide preparers and users of advanced visualizations has been argued elsewhere (Orland 1992; Sheppard 2001), including a Code of Ethics for users. While it is necessary to review ethical principles and procedures for landscape visualization, the focus here is on the technologies or tools themselves, and the implications of ethical issues for design of the next generation of visualization media. This chapter is also primarily concerned with the use of visualizations in practice, rather than as experimental research tools: in particular, the use of largely predictive landscape visualizations in the fields of design, planning, public involvement, resource management decision making, and general awareness building on related issues. In this context, it is important to consider both the more conventional use of visualizations as discrete image sets used as a component in the design and approval process, and more 'intelligent' visualizations built into larger, multi-criteria decision-support systems and educational devices.

Certain terminology used in this chapter should be defined. The term 'visualization systems' refers to the overall combination of technologies and the support infrastructure available to assist the preparer of the visualization (such as manuals, procedures, data, training, etc.) in implementing the technologies. The term 'visualization tools' refers to the 'crystal ball' itself: the actual technologies and media used to prepare and present landscape visualizations, including hardware, software, and the interfaces used to display and interact with the user/viewer.

This chapter summarizes ethical principles for the use of landscape visualization systems, and advances appropriate characteristics associated with quality of visualizations (Question 1), although it should be said at the outset that many of the supporting assumptions have not been rigorously tested through research. There follows a brief review of some of the evidence for how well actual visualizations are performing, addressing, in particular, the question of whether existing tools are good enough (Question 2). The chapter concludes by suggesting ways in which visualization tools could be designed to build in more ethical usage (Question 3).

Defining a 'good' landscape visualization

What does 'good' or 'better visualization' mean, in the context of ethical practice? 'Better' is often interpreted to mean faster, higher tech, more realistic, more user-friendly. However, these aspects have as much to do with efficiency, profit, image, popular demand and selling a product or a design, as they do with more meaningful factors for protecting the public interest: e.g. safer and more informed decisions, defensibility and other ethical considerations. Ethics refers to moral principles which distinguish between right and wrong; in professional practice, this usually means conforming to a recognized standard or rules of conduct. Validity generally refers to whether an instrument or finding is sound, defensible and well-grounded or appropriate to the issue at hand. Reliability refers to consistency in repeated applications. Both these concepts are central to the ethical use of landscape visualizations.

Validity of landscape visualization

Scientific definitions of validity identify several forms of validity, including the following.

- *Content or face validity*: Are we measuring what we think we are measuring? Content validity addresses whether the content of an instrument (in this case visualization) matches the desired objectives of its use and makes sense in its context (Weller and Romney 1988). For example, if we want to judge the visual impact of a proposed building, does the simulation clearly show important views of the correct building design proposal?
- *Criterion-related or predictive validity*: Do the measurements derived from use of the instrument satisfactorily predict the external measurement (or criterion) in question? Thus, if the intent is to predict human behaviour, there should be a close relationship between the predicted versus observed behaviour (Weller and Romney 1988). With visualizations, for example, if the intent is to predict the attractiveness of a design, it is important that responses to the visualized design are similar to those

Understanding visualization

obtained from the actual design once built; this has been termed 'response equivalence' (Craik *et al.* 1980) or 'representational validity' (Daniel and Meitner 2001).

- *Construct validity*: Are there consistent results among various approaches or instruments that are supposed to measure the same thing? Weller and Romney (1988) advise triangulation of methodological approaches in order to discover true relationships and build confidence in the findings. Validating responses to a proposed project, for example, requires comparison with information external to the original findings obtained with a particular visualization; this might require multiple forms of visualization and other sources such as testimony from viewers of similar projects built elsewhere. Similarly, the validity of the construct that proposed large clearcuts would be ugly is supported if different visualization tools lead to similar negative aesthetic responses. Significant differences in responses obtained with different visualization media would throw doubt not only on the construct or conclusion, but also on the validity of at least some of the visualization tools used.

Pragmatically, there are in fact a number of *dimensions* of landscape visualization which relate to the scientific definitions of validity just described. Realism has been defined in terms of *actual realism* – response equivalence or lack of bias in responses between simulated and real environments, as described above – and *apparent realism* (Figure 5.1) – the degree to which the simulation appears to look like the real world when

5.1 Apparent realism: example of a highly photo-realistic landscape visualization of unknown accuracy or actual realism

5.2 Pictorial inaccuracy: comparison of (left) a visualization of a proposed power plant versus (right) a site photograph of the actual project in operation, showing substantial visual differences in content and appearance

judged on the basis of the image alone (Sheppard 1982; Lange 2001). The latter can be important in assuring credibility, which can be considered an aspect of validity under the general definition of the term. *Accuracy* is another concept strongly related to realism: it addresses the issue of truthfulness or fidelity of the visualization imagery to the actual or expected appearance of the landscape in question (Appleyard 1977; Sheppard 1986). It is analogous to response equivalence in terms of before/after comparison, although it is directly measurable in terms of objective image qualities (Figure 5.2), rather than as mediated through observer responses. It relates also to the concept of 'ecological validity' described by Palmer *et al.* (1995), referring to the assurance that the environments visualized are ecologically feasible.

Accuracy and apparent realism have traditionally been defined as similarity in pictorial elements between a static simulated view and a static real-world view (actual or imagined); see, for example, Palmer *et al.* (1995) and Lange (2001). Conceptually, it can be extended to cover aspects of animated visualization in terms of accuracy or realism in movement in relation to the real world (e.g. speed and direction of simulated travel), or in interactivity (e.g. control over direction and changeability of viewing) (Bishop 2001). Another related dimension of validity is *representativeness*, concerning the choice of viewing position; an animation which flies the viewer around a tall building may be strictly accurate to the design and photorealistic, but of questionable validity for evaluating public reaction, since no one could recreate this viewing experience in the real world.

Visual clarity is also important in communicating effectively the appropriate message; a poorly reproduced or confusing image poses a direct threat to face validity, since we cannot be sure what the viewer is responding to.

It is important to realize that these dimensions are closely linked but can operate independently: a highly realistic-looking image may in fact be very inaccurate or even show a completely fantastic scene with no ecological validity; and a visualization that is very inaccurate on many attributes may still lead to the right response to some questions. When we address the issue of validity, we must therefore ask: 'valid for what?'. Different response types (e.g. cognitive, affective, evaluative) may lead to different levels of response equivalence from the same visualization. Is response equivalence always needed? If the cognitive response is the real need, then response equivalence, normally the 'holy grail' of visualization validity, may in fact not always be desirable. If the intent of the visualization is to explain underlying truths or clarify spatial complexities (e.g. navigating a path through a complex building), then a highly realistic experiential simulation (Appleyard 1977) that is as confusing or opaque as the real world may fail in its purpose, although highlighting a real design problem.

The various dimensions of visualization validity can be grouped into two main camps, based on when they are applicable in the life cycle of a project. Many are impossible to measure directly until after the proposed project is built and in use (although the simplistic paradigm of the 'before/after' image pair fails to recognize the continuously changing environment after project construction). The other group of dimensions is directly measurable prior to construction, during the planning process when visualizations are actively used, and act as proxies for the ultimate tests of validity. These validity indicators, which form the basis for the principles of visualization discussed below, reflect the precautionary principle in the absence of comprehensive hard research results on predictors of post-construction validity. For example, accuracy can be defined in terms of truthfulness to the best data available at the time, or to the details of the proposed design. By analogy, a map of the world showing the Suez Canal in the wrong place is clearly wrong, even if it never leads to navigational problems in actuality.

Reliability of landscape visualization

Scientific definitions of reliability refer to internal consistency of methods: 'the degree to which an instrument ... can retrieve the same answers when applied under similar conditions' (Weller and Romney 1988, p. 70). Palmer and Hoffman (2001) define reliability as 'the dependability or consistency of something that is done repeatedly'. For example, if the same software and project data are used by different operators and lead to very different visualizations and responses, this would not be considered reliable.

In practice, reliability requires consistency in the visualization process, and consistency in objective: opponents to a project may create a very different set of visualizations than the applicant, using the same datasets, but with different motivations. Reliability of visualizations can potentially be affected by many diverse factors, such as data quality, operator procedure and skill level, deliberate bias or stakeholder influence, 'bugs' or data incompatibilities in the software, etc. Most of these factors also threaten validity.

Quality thresholds for landscape visualization

So, how do we know when a visualization is good enough? This requires setting standards or thresholds along the various quality dimensions or predictive indicators. Some of these thresholds have been tentatively identified through perceptual research, particularly in terms of media technology: examples include levels of abstraction in photomontages (Sheppard 1982); image resolution (Bishop and Leahy 1989), and colour accuracy (Daniel *et al.* 1997). Key thresholds also crop up in practice and applied research: in work with stakeholders in the Slocan Valley of British Columbia, for example, we have encountered significant issues of public credibility on details such as the age at which slight colour and textural shifts occur in photographic tree models representing second-growth forests. However, most of the myriad possible thresholds to consider in realistic landscape visualization have not been established. Pending further research and testing, the next two sections suggest some general normative thresholds in the definition of principles and accompanying code of ethics. More specific thresholds of concern may have to be derived on a context-specific basis, perhaps through advance consultation with the agencies and affected stakeholders. The general adequacy of current visualization tools is discussed below.

Principles for landscape visualization

The real need for visualization is to provide better means of communication and to support more informed decisions. Sheppard (1989) identified three fundamental objectives for landscape visualization in practice: to convey understanding of the proposed project; to evoke unbiased responses to the proposed project; and to demonstrate credibility of the visualizations themselves. All of these objectives relate directly to validity and reliability.

In order to meet these fundamental objectives, in the absence of recognized or formalized standards for landscape visualization, the following six principles (adapted from Sheppard 2001) are proposed, cutting across a range of media, approaches, and uses of landscape visualization:

1. accuracy
2. representativeness
3. visual clarity
4. interest
5. legitimacy
6. access to visual information.

Definitions for these principles are provided in Box 5.1. These principles provide suggested guidance on the quality of visualizations at the pre-construction stage. Principles 1–3 relate directly to issues of content validity, as discussed above. Principle 4 primarily addresses utility, by ensuring that viewers or users remain engaged in the process, although lack of interest could also influence the validity of responses through viewer fatigue. Principle 5 addresses credibility and acceptability or satisfaction of the viewers/users with regard to the quality of the visualization and procedures followed, which relates directly to validity and reliability. Principle 6 addresses the concept of equity (equal access for various stakeholders and members of the public), which can also be seen as a validity issue. These principles are intended as proxies for the ultimate tests of validity and reliability, which can only be conducted after projects have been implemented as designed.

Principles such as these provide the basis for interim procedural guidance for ethical landscape visualization, briefly described next.

An interim code of ethics

Beyond the choice of software or medium used, many factors influence visualization quality: for example, content choices, viewpoint choices, presentation mode, etc. Orland *et al.* (2001), among others, describe the ideal technological components of a decision-support system integrated with visualization, but lament the lack of guidance for these emerging systems.

With steadily increasing access to more user-friendly software, the lack of training or guidance in the use of visualization poses a significant threat to valid public processes. The heavy reliance on imagery to sell market-driven products makes it inevitable that deliberate distortion for commercial purposes will be attempted in planning, too. One of the most urgent needs then is to develop widely recognized ethical guidelines for landscape visualization. A code of ethics should establish clear guidelines on appropriate approaches to preparing and presenting visualizations of future landscapes, and monitor their use in order to adapt and strengthen the guidance itself. A code would provide a means to determine whether a given visual representation/process has met the recommended procedures or minimum standards of the profession. At the same time, it would provide a defensible basis for informing clients, approval agencies and the public that no visualization can be completely accurate or fully representative of a built project. It has to allow for uncertainty and flexibility in this world of rapidly changing technology and diverse landscape projects.

Pending comprehensive findings from the considerable body of research which is needed on this subject, an Interim Code of Ethics has been proposed for consideration, testing, and amendment by other researchers and users (Sheppard 2001), and is presented in adapted form in Box 5.1. It attempts to translate the six principles into more practical approaches and general procedures intended to provide reasonable and feasible safeguards for validity, reliability, and other aspects of visualization quality. More work needs to be done on this to develop better definitions of 'reasonable' and 'appropriate' methods.

Validity, reliability and ethics in visualization

Box 5.1 An interim code of ethics

Proposed interim code of ethics for landscape visualization – version 4

Purpose of landscape visualization
Professional preparers and presenters of realistic landscape visualizations are responsible for promoting full understanding of proposed landscape changes; providing an honest and neutral visual representation of the expected landscape, by seeking to avoid bias in responses (as compared with responses to the actual project); and demonstrating the legitimacy of the visualization process.

General principles
Preparers and presenters of landscape visualizations should adhere to the following general principles:

- Accuracy: realistic visualizations should simulate the actual or expected appearance of the landscape as closely as possible (at least for those aspects of the landscape being considered); visualizations should be truthful to the data available at the time.
- Representativeness: visualizations should represent the typical or important range of views, conditions, and time-frames in the landscape which would be experienced with the actual project, and provide viewers with a range of viewing conditions (including a visualization of typical worst-case conditions at a minimum).
- Visual clarity: the details, components, and overall content of the visualization should be clearly communicated.
- Interest: the visualization should engage and hold the interest of the audience, without seeking to entertain or 'dazzle' the audience.
- Legitimacy: the visualization should be defensible by following a consistent and documented procedure, by making the simulation process and assumptions transparent to the viewer, by clearly describing the expected level of accuracy and uncertainty, and by avoiding obvious errors and omissions in the imagery.
- Access to visual information: visualizations (and associated information) which are consistent with the above principles should be made readily accessible to the public via a variety of formats and communication channels.

Code of ethical conduct
The use of landscape visualizations should be appropriate to the stage of development of the project under consideration, to the landscape being shown, to the types of decisions being made or questions being addressed, to the audience observing the visualizations, to the setting in which the presentation is being made, and to the experience level of the preparer. In general, preparers and presenters of landscape visualization should:

- demonstrate an appropriate level of qualifications and experience
- use visualization tools and media (more than one if possible) that are appropriate for the purpose
- choose the appropriate level(s) of realism
- identify, collect, and document supporting visual data available for or used in the visualization process; conduct an on-site visual analysis to determine important issues and views
- seek community input on viewpoints and landscape issues to address in the visualizations
- provide the viewer with a reasonable choice of viewpoints, view directions, view angles, viewing conditions, and time-frames appropriate to the area being visualized
- estimate and disclose the expected degree of error and uncertainty, indicating areas and possible visual consequences of the uncertainties
- use more than one appropriate presentation mode and means of access for the affected public
- present important non-visual information at the same time as the visual presentation, using a neutral delivery
- avoid the use or the appearance of 'sales' techniques or special effects
- avoid seeking a particular response from the audience
- provide information describing how the visualization process was conducted and key assumptions/decisions taken
- record responses to visualizations as feedback for future efforts
- conduct and document post-construction evaluations to assess accuracy of visualizations or changes in project design/construction/use.

An evolving version of the Interim Code of Ethics has been posted on the Collaborative for Advanced Landscape Planning (CALP) University of British Columbia web site (www.calp.forestry.ubc.ca) in an attempt to foster comment, discussion and improvements.

Why do we need a better crystal ball?

How good are visualization methods already? Is the crystal ball we are using good enough? What might we gain in quality if we move to still more advanced technology?

General evidence of the quality of current forms of landscape visualization comes from three sources:

- research findings based on empirical evidence: this is somewhat limited, especially in validating the newer, more advanced visualization tools;
- anecdotal evidence from practice and observation of trends, which are rather poorly documented for the most part; and
- normative/deductive reasoning from theories and assumptions, in order to fill the gaps left from the other two sources.

The following discussion focuses on the quality of tools and media, although it is often hard to distinguish between issues arising from the technology versus the process of creating visualizations.

Arguments for not needing a better crystal ball

There would appear to be several arguments on grounds of quality for not needing better visualization technologies, at least, not yet. From a practical standpoint, existing technology provides products which already 'look good'. Off-the-shelf programs such as Visual Nature Studio (the software formerly known as World Construction Set), Bryce and World Builder can create highly realistic and totally convincing landscape imagery (see Figure 5.1). Current simulations are often claimed to be 'data driven', being populated with GIS-based, georeferenced landscape data and digital terrain models: are they not then defensible already? World Construction Set (WCS) images used in forestry in British Columbia are often readily accepted by observers as a solid basis for judgment of aesthetics and other factors, although the visual images themselves can vary considerably from those created by other realistic visualization methods (Figure 5.3).

There is some limited scientific evidence that certain visualization tools, some of which have been around for a while, can deliver valid responses for various types of questions. Over 20 years ago, Craik *et al.* (1980) proved that they could replicate people's responses to the real world through an animated video tour of a realistic scale model.

5.3 Alternative media: comparison of visualizations of a proposed timber harvesting plan created by (top) photo-imaging (guided by a three-dimensional canopy model) versus (below) World Construction Set.

Oh (1994) found that image-processed simulations correlated very well with photographs in nearly all of the visual response measures used. However, much research remains to be carried out before we can conclude what works and what does not work with newer visualization tools such as WCS or CommunityViz, for example. Wherrett (2001) conducted a rare comparative test of some forms of commercial three-dimensional landscape visualization, and found poor correlations in responses between modelling and existing site photographs, although it is unclear whether this resulted exclusively from the media employed. At the Collaborative for Advanced Landscape Planning (CALP) we have begun to compare responses to forest landscape scenes represented in WCS with corresponding site photographs and other forms of visualization. Lewis (2000), for example, has shown that WCS visualizations can lead to more reliable cognitive responses than two-dimensional GIS maps with an aboriginal community in British Columbia. These examples notwithstanding, there is a strong argument that what we need most is not new technology, but more research on the methods we already have.

It also appears that technology may not be the most limiting factor in visualization quality currently; other factors such as the type of project (Bishop and Leahy 1989), human error, or the process itself (Sheppard 1989) may be more urgent priorities for investigation. Other scientific arguments for not needing a better crystal ball are based on the lack of data to populate the current high realism techniques, let alone more sophisticated systems (Orland and Uusitalo 2001). With forest visualization, for example, data on visual attributes of existing landscape components or future forest stand conditions are not systematically available as part of normal forest inventory data.

Finally, some emerging forms of visualization may be positively dangerous in an objective decision-making context. Users may see more elaborate technologies with real time, big screens, massive datasets, etc., as overwhelming or suspect, as overly-persuasive 'bells and whistles'. Orland et al. (2001) question whether interactive immersive technologies are intrinsically antithetical to objective decision making, since they reduce detachment and can actively invoke emotional responses.

Arguments for needing a better crystal ball

Practical arguments for improving visualization tools respond to the apparent public hunger for new 'high-tech' features, such as higher realism, speed and interactivity. Forest managers and scientists talk enthusiastically of their desire to glide over the future virtual forest and dive down to look at the trees at close range, all of which requires computing resources not yet affordable or widely available in commercial systems if any level of realism is required. There is genuine excitement and anticipation here of improved usefulness, although the specific benefits of conducting such an exercise often are not clearly articulated.

Social and scientific arguments for more advanced technologies rest upon evidence of problems with current media or anticipated problems and demands with emerging visualization trends. Unfortunately, while there are many anecdotes and widespread skepticism concerning visualization validity in particular (e.g. McQuillan 1998; Sheppard 2000; Luymes 2001; Orland *et al.* 2001), as noted above, the hard data is very limited. A few authors have found bias in some response dimensions with some media in comparison to the real landscape or photographic surrogates (e.g. Sheppard 1982; Bishop and Leahy 1989; Bergen *et al.* 1995; and Wherrett 2001). This indicates problems with at least some previous visualization media, although there has not been sufficient research to establish systematic relationships, for example between accuracy and response equivalence.

We are therefore left with largely normative or deductive analysis. Despite today's advanced computer visualization programs, many of the past limitations on visualization which pose theoretical threats to validity still apply in current practice. Some examples are listed below.

- Use of fixed, static and limited images: the confines of the rectangular frame and the small number of still images are still largely with us, offering a tiny and potentially misleading slice of visual information, with little or no choice of viewing alternatives. Meitner and Daniel (1997) have established the increased benefit of permitting 360° view choice through interactive Quick Time Virtual Reality techniques, compared with fixed slides, when replicating on-site scenic beauty judgments.
- Use of small image sizes and a narrow view angle, providing no panoramic or peripheral vision (Figure 5.4): this divorces projects from their context, and could result in misleading impressions of relative scale and dominance of objects in the field of view (Danahy 2001a; see also the section 'The visualization of windfarms', p. 184).
- Reliance on a single snapshot in time, with no sense of change over time or dynamic thresholds, e.g. when tree growth begins to block critical views. Dynamic modelling systems such as Smart Forest-II (Orland 1997), or as described in Chapter 6, are beginning to break down these barriers. In practice, however, predictive simulations cannot be said to be data-driven: most of the data used in preparing realistic visualizations is limited existing data (e.g. timber inventory data, existing photographs, satellite imagery) and, inevitably, the preparer has to guess how visualized appearance should look and change over time.
- Restrictions on the area modelled or real-time movement through extensively vegetated/forested landscapes, due to the incredibly large number of polygons to be rendered.

These would seem to be areas where technological improvements could be beneficial.

5.4 Differences in field of view: comparison of (left) landscape photographs with conventional aspect ratio versus (bottom) panoramic imagery, showing differences in content and context of back-country recreational views

As landscape visualization moves beyond issues of aesthetics and becomes increasingly linked to broader environmental modelling and decision-support systems, new opportunities and risks may call for technological advances. Making valid and meaningful decisions that balance the complex arrays of data and issues, in an increasingly public forum, will require easier forms of retrieval and viewing, connecting directly with non-visual data through the visualization medium. Current programs like WCS routinely provide detailed realism for forestry in British Columbia, but cannot show other aspects of the forest which may be equally important, such as soil productivity or habitat sensitivity. There is also the clear risk of bias through the public reviewing limited, selective packets of information via small or distorted visualization 'windows'. The general trend towards transparency in decision making and more participatory public processes may require more hands-on or transparent visualization forms, with more access for and control by the viewer, as in new programs such as CommunityViz (see the section 'Visualization in support of public participation', p. 251).

However, as mentioned above, potential threats to validity associated with such emerging visualization tools may need to be overcome. Luymes (2001) and Orland and Uusitalo (2001) fear that the increasing immersive realism, sophistication and implied authority of emerging visualization methods may make it difficult to suspend disbelief in the displays, potentially misleading viewers as to the certainty or accuracy of the data. It appears, therefore, that more sophisticated technology does not necessarily mean increased transparency. It is also possible that the power of immersive displays may exaggerate the importance of visual phenomena over non-visual ecological conditions. Such validity issues could be compounded by making potentially misleading imagery available to millions over the World Wide Web.

Are there, perhaps, safeguards against inappropriate use of visualizations in these expanding contexts, which can be built into the new technology from the outset? This idea is explored below. Ultimately, however, debating whether further technological development is necessary is futile, since rapid change appears inevitable. Fundamentally, the key issue is how to best influence that development, in order to overcome the visualization problems of today and tomorrow.

Ethical design for the new crystal ball

As has been shown, many of the drawbacks with current visualization systems reflect a combination of technology limitations and user/process limitations. This section explores how visualization tools might be improved by building aspects of the support infrastructure into the technology, bridging the gap between a code of ethics and technical specifications. Several authors (e.g. Orland *et al.* 2001; Orland and Uusitalo 2001; and Danahy 2001b) have reviewed technological requirements of emerging virtual reality tools and implications for design and decision making. This analysis considers the design of tools more narrowly, from the perspective of validity, reliability, and ethical considerations. It extrapolates from the foregoing analysis and draws on design evaluation criteria documented by Cavens (2002) for interactive landscape visualization prototypes.

There would appear to be two conceptual design approaches to providing safeguards or limits on threats to visualization quality:

1 more prescriptive approaches which guide or drive the presentation of visualization material according to established principles or standards; and
2 more flexible and interactive approaches which give much greater control over visualization information to the user/viewer.

Both of these approaches provide greater access to and transparency of visual imagery and underlying metadata than is normal with current tools, although in different ways and to differing degrees.

The more prescriptive approach involves a standard format or viewer interface which explicitly relates the visualization process to the final products. It would foster compliance with principles, a code of ethics, or detailed protocols by 'boxing in' (to a greater or lesser degree) the documentation and delivery of completed visualization imagery. Examples might include formats that document the rationale for viewing conditions selected, structure the display of alternative levels of realism and record other metadata. This could take a 'soft' form, offering templates for supplying the key information, or a 'hard' form requiring input of at least some key data in order to enable the viewing/presentation program to run. This would provide a measure of consistency in the delivery of information to the viewers, force visualization preparers to think about a range of key validity/reliability issues, and provide some evidence that recognized procedures have been followed. The viewer interface would work with any visualization tools or software, and could be used either as a presentation device for the presenter or an access window to preset visualizations for users/viewers on their own time, via the Internet or computer terminal.

The second approach, rather than attempting to prescribe how visualization material should be prepared and presented, would give the viewer much more choice and control over what they see, and freedom to roam within the visualization dataset. This has more direct implications for the design of the visualization tool and/or process itself, since it would require a much larger volume of visual information and possibly a means of accessing or manipulating the visualization directly by the user/viewer. Various levels of user/viewer choice and control can be imagined.

1. The simplest level would involve viewing a wide range of 'pre-canned' visual information, such as a multitude of possible (but predefined) viewpoints or view paths, where the user/viewer can select which route to take through a three-dimensional landscape model, where and when to stop, what direction to look at in panoramic views, etc. Precedents for this approach include: the ground-breaking visualization set published on CD for the Guanella Pass proposed road improvements in Colorado (Taylor and McDaniel 1997), which included 3D Studio animations of alternative routes, QTVR panoramic images, and multiple static photosimulations along the highway corridor (Figure 5.5); and

5.5 Sample frame from the Guanella Pass Proposed Road Improvements Project CD ROM, showing a drive-through video-animation in a user-selected driving sequence of the proposed highway improvements (image reproduced in the colour plate section)

Understanding visualization

time-lapse visualizations formatted in Macromedia comparing alternative forest harvesting scenarios from multiple viewpoints in the Slocan Valley of British Columbia (see http://www.calp.forestry.ubc.ca/projects_Arrow.htm). In theory, key metadata could be linked to these user-defined views in real time.

2   An intermediate level might involve more interactivity, with the user/viewer defining their own view path through a predefined three-dimensional landscape model, and querying predefined packages of information directly via the screen, e.g. material palettes for texture maps used in the imagery (Cavens 2002). Early examples of these would include: academic prototypes such as the Java forest landscape viewer developed at CALP in 1999 (Figure 5.6), which allowed user-defined overlays and transitions between site photographs to WCS models to underlying terrain models; and emerging commercial systems such as Viewscape 3D's OpenGL Ecoviewer for viewing proposed forest landscape plans interactively.

3   The most sophisticated tool would allow the user/viewer to manipulate the landscape model itself, re-rendering the landscape or project conditions in real time as well as choosing their own view path, viewing conditions, or time frames. Precedents for such a system include academic prototypes such as Forester

5.6 Making visualizations transparent – a simple Java application interface designed to display simultaneously both the modelled image and the underlying data: (top left) three-dimensional forest canopy wireframe model; (top right) existing site photograph; (right) billboarded tree templates (images reproduced in the colour plate section)

(Chapter 6, the section 'Planning, communicating, designing and decision making for large scale landscapes', p. 120), the PA suite (Chapter 7, 'Helping rural communities envision their future', p. 145), and current commercial systems such as CommunityViz (see Chapter 10, 'Visualization in support of public participation', p. 251) which allow user-defined OpenGL real time movement and direct linkages to numerous non-visual data-modelling parameters. Such systems require sophisticated interfaces, with the potential for various interface tools allowing non-experts to design or adapt proposed projects for rapid visualization, to view any stage in the preparation of visualizations, or to replicate the visualizations prepared by experts and thus assess their reliability directly.

The technological developments described in this second approach promise to democratize the planning process to an unprecedented degree. Providing so much choice and freedom to roam within the visualization material would reduce the risk of systematic bias from a presenter or preparer, and allows more direct and individualized estimates of (pre-construction) validity and reliability. However, there is nothing to say that self-choice would automatically lead to the appropriate information being seen or found, and it could lead to more biased or confusing arrays of visualizations for public decision making. Some combination of the first and second approaches might ultimately lead to the greatest validity, in providing an ethical structure to evaluate the information while preserving public freedom to explore the visualization process and design options.

There remain, however, serious questions about the overall effect on validity of certain technological advances described above. The value of increased user/viewer interactivity could be important if it enables better decisions through increased user engagement and understanding, and by reducing skepticism or resolving reliability concerns. However, do we have the public participatory mechanisms and guiding policies to deal with the speed of interactivity? In a public setting, can we really trust snap decisions made rapidly in computer model or interactive display? The use of new interactive tools such as CommunityViz in public settings needs to be monitored to learn about the risks and benefits.

The role of immersive display technology also requires more research. The mode of presentation of any visualization material is clearly very important (e.g. Danahy 2001b). The principle of response equivalence would seem to support more immersive displays (Figure 5.7) which recreate the correct image size and field of view (Sheppard 2001) of the real world view; yet, Orland *et al*. (2001) have pointed out that this kind of immersive engagement conflicts with the detached analytical view associated with resource decision making.

The utility, feasibility and cost implications of using the more advanced technological improvements described here, although not the

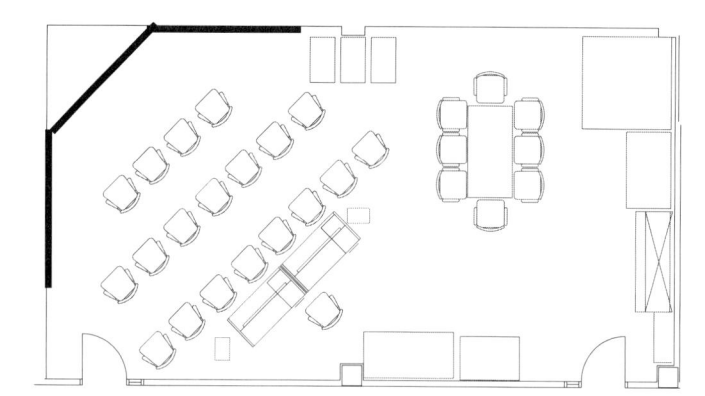

5.7 Semi-immersive displays: facilities such as the Landscape Immersion Lab at UBC can provide panoramic views at life size for small groups of people

focus of this chapter, are not trivial. Trade-offs between different dimensions of realism or functionality supporting these tools will need to be made, at least in the short term; Figure 5.8 shows how current visualization technologies can be classified on multiple dimensions relevant to visualization quality, in this case relating levels of realism to levels of interactivity. Clearly, however, it would be feasible and beneficial to introduce some of the simpler technological improvements, for example building in simple devices such as accessible metadata files and static viewers to demonstrate accuracy of photorealistic visualizations against site photographs (see Figure 5.6).

We cannot rely solely on improved technology to deliver quality in landscape visualization; we need some supporting infrastructure to provide guidance for crystal ball gazers. We can begin to lay out the precautionary principles and interim ethical procedures to guide predictive work in landscape visualization, while we encourage the much-needed research to test, adapt and validate these principles and codes themselves.

Nonetheless, there is a case for building a better crystal ball: the design of landscape visualization tools should be driven in part by the needs of validity and reliability, integrating ethical guidance into structures and interfaces of the technology itself. Developing and using tools such as standard viewer/metadata formats and queriable dynamic simulations should raise expectations for the defensibility and transparency of landscape visualizations, and help to demonstrate just how much confidence should be invested in their widening application to design and resource decision making.

Acknowledgements

The author would like to acknowledge the contribution of graduate student researchers Jon Salter and Duncan Cavens to ideas and visual-

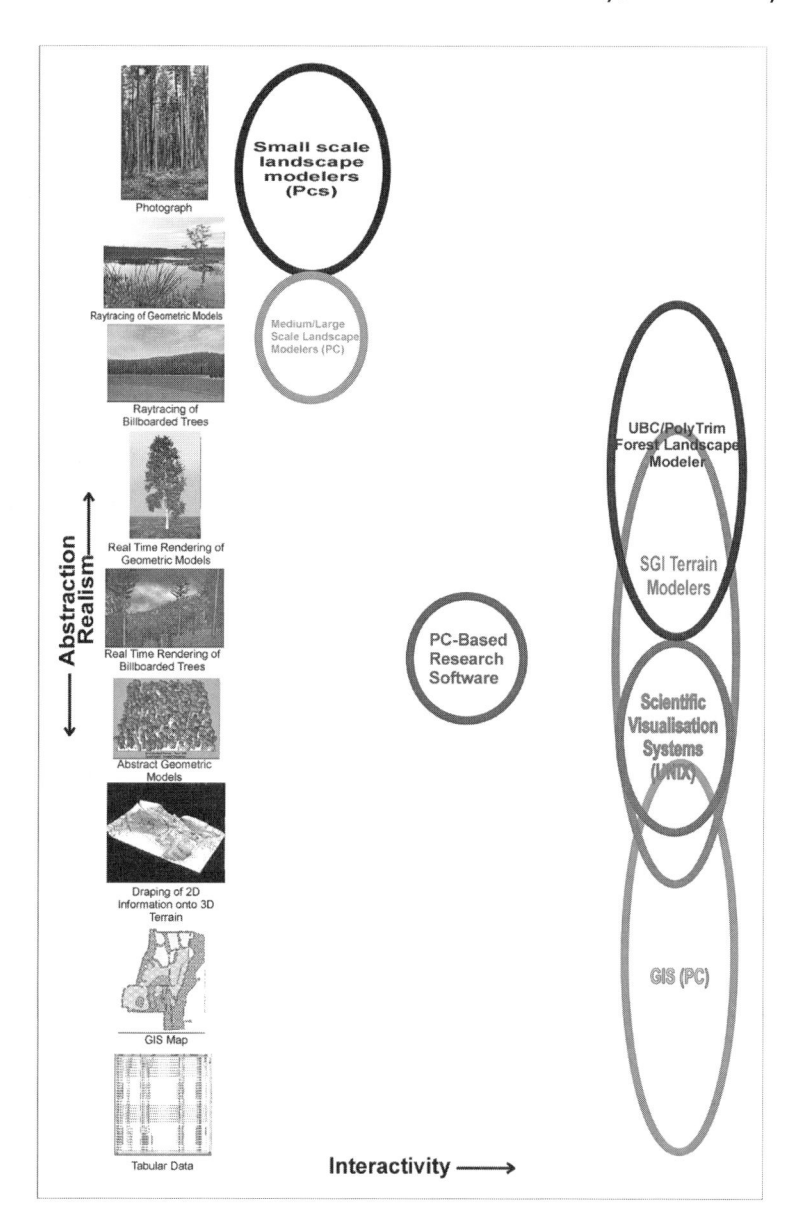

Validity, reliability and ethics in visualization

5.8 Classification of visualization tools arrayed on a graph of photorealism against interactivity: more sophisticated technologies are moving towards the upper right corner

ization prototypes discussed in this chapter. Thanks also go to Dr Michael Meitner who has been instrumental in the discussion and evolution of concepts addressed here. Caitlin Akai provided invaluable support in the preparation and editing of the manuscript.

Part 2

Applications

CHAPTER 6

APPLICATIONS IN THE FOREST LANDSCAPE

INTRODUCTION

Duncan Cavens

There is a long history of using computer visualization in forestry, starting with early research in the 1970s (Kojima and Wagar 1972; Myklestad and Wagar 1977). This work, along with most subsequent development in the area, was inspired by the forestry industry's long-standing need to manage the visual impact of its traditional forest harvesting practices.

The impact of these practices on visual quality has long been recognized as one of the most difficult issues in the forestry industry. Modern industrial harvesting techniques, in particular clear-cutting, are widely perceived as being unsustainable and visually disturbing. In parts of the world (particularly the western coast of North America), this concern is so acute that it has resulted in protests and mass arrests, and has a significant impact on the economic viability of the forestry industry.

As a result, there has been considerable research effort into understanding how the public reacts to forest harvesting (see Magill 1992; BCMoF 1996) and how to minimize the public's negative reaction. Initially, the public concern resulted in guidelines and regulations (see for example USDA Forest Service 1995; BCMoF 1997) that were developed to minimize the visual impact of forest harvesting operations. However, as these regulations usually rely on numerical approximations (such as percentage of a view altered) for concepts that are not easily quantified (such as public acceptability), their success has been mixed. These prescriptive regulations often result in substantially less timber being available for harvest, while not necessarily meeting the public's expectations related to visual quality (Picard and Sheppard 2001). As a result, there is increasing interest in developing site-specific solutions to visual quality concerns, using alternative harvesting techniques. Visualization tools have the potential to greatly assist this approach, as forest managers are able to iteratively test different management prescriptions, and present them to non-foresters for feedback (Figure 6.1).

While visualizations can help to discover more palatable alternative prescriptions, they can also assist forest managers in justifying a particular management decision. Like other areas of landscape planning, there has been an increasing tendency to include the concerned public directly in the forestry planning process. However, it is particularly difficult to convey the complexity of forest decision making, which is a classic example of a complex long-term ('rotations' between harvests can be up to 250 years) trade-off between economic, ecological and societal concerns. Visualization is a powerful tool that allows everyone to 'see' how a landscape will change over a very long time span, and not simply focus on the immediate impact of a harvesting event.

As McGaughey (1997) describes, a variety of techniques have been used in forest visualization, ranging from image-based techniques to full geometric modelling of a landscape by representing each individual tree. As computer hardware and graphics algorithms have improved, increasingly the standard has become representing an entire landscape with highly realistic image-based representations of millions of stems. Over the past decade, techniques and commercially available software have evolved considerably, to the point where highly realistic simulations have been used routinely in the forest approvals process in different jurisdictions in North America.

While it has become easier to create high quality visualizations, many questions still exist about how to create images that are accurate and scientifically defensible, particularly when visualizing the changes in a forest landscape over a time-scale of decades. The visualizations

6.1 Current techniques allow high quality visualizations of very large scale landscapes

currently being produced by industry, while visually impressive, generally contain very little forestry-related information: they represent long-distance views and reflect conditions immediately before and after a single harvest event. More research and development is required to realize the full potential of forestry visualization. The research presented in this chapter describes the ongoing effort to extend the capabilities of forest visualization to more intimate views; to couple realistic models of forest regrowth to visualizations in order to represent long-term forest dynamics accurately; and to evaluate when and how to integrate the visualizations into public processes.

Specifically, Brian Orland, in his contribution ('Calibrating' images to more accurately represent future landscape conditions in forestry, p. 104), describes how difficult it is to calibrate images to existing and future site conditions, particularly when the emphasis is on visualizations that represent the middle ground. In their section (Studying the acceptability of forest management practices using visual simulation of forest regrowth, p. 112) Ian Bishop *et al.* describe, using a recent example from south-eastern Australia, a specific technique for representing forest dynamics in the middle ground, and demonstrate how much detail and data is required. Salter *et al.* (Planning, communicating, designing and decision making for large scale landscapes, p. 120) use an example from the interior of British Columbia, Canada to examine the benefits of integrating a visualization system with complex forest stand, landscape and biodiversity models for use in a highly contentious stakeholder process. Finally, Tyrväinen and Uusitalo (The role of landscape simulators in forestry: a Finnish perspective, p. 125) describe the current state-of-the-art of forest visualization in Finland, and describe how three different forest visualization systems currently in use in Finland can be used for a wide variety of purposes.

'CALIBRATING' IMAGES TO MORE ACCURATELY REPRESENT FUTURE LANDSCAPE CONDITIONS IN FORESTRY

Brian Orland

The concept of 'calibrated' images was born out of the need to match specific levels or combinations of changes to images that could communicate some of the complex issues facing natural resource managers. Visualizing future forest conditions challenges our ability to project changes with certainty. No part of our image of the future can be assumed to be static, as we know that both subject and context for our representations are constantly growing and changing. A calibrated image does not pretend to represent the future but instead seeks to represent the best numerical or expert judgment about that future. This is not merely a semantic difference but signals an important division of responsibilities. One responsibility is to ensure that the visual imagery that is eventually used matches the numerical data that are provided; the other is to ensure that those data have the support of resource experts as to the validity of their representation of the projected future.

Close in scale and time

Early uses of computer-based image editing applied to changes in vegetation included Orland *et al.*'s (1992) study, where images of residential properties were edited to include smaller, larger, or no trees in front of them in order to ascertain the contribution of trees to perceived property value. In this elementary example the calibration of the image to the experimental question is simple to validate. At the close scale, as in this example, trees are managed and comprehended as specific individual entities and there are sufficiently few of them that case-by-case treatment of their visual appearance is feasible. While change over time may be a serious concern, images can readily represent different stages of development by reference to other trees of known age.

Distant in scale and time

A key factor in our ability to calibrate images to represent the changing natural world is the nature of the data we collect about that world. At the close scale, as described above, a tree may be known by species, age, condition and a number of measures of size. At the middle and large scales the calibration of images to known or anticipated ground conditions becomes more challenging.

Trees, woodland or forest at the scale of a city park or a campsite in a public forest will rarely be inventoried at the level of the individual tree. An inventory will be based on samples, statistically extended to represent the full forested area. Sampling approaches are by necessity limited in their scope. Choices might be made to sample the most easily measured elements – such as the trees over 75 mm diameter – or those of high commercial, ecological or cultural value. Whatever happens with the tree inventory, almost inevitably the ground or middle-level grasses, forbs and woody materials will receive a lesser level of inventory. There are many more of them, they show more variety and, with rare exceptions, lack the 'charisma' of the trees that leads to them being carefully measured and recorded. Incomplete, over-generalized or missing data each present challenges to the image-creation process.

Forest at the scale of a regional landscape, such as a watershed or viewshed, may receive detailed inventory in its many parts, as a summary of sampled data. Forest type descriptions – Douglas Fir-White Fir, Ponderosa Pine, Birch-Spruce – encompass the types of ground and small woody material that accompany the trees. Thus, while scale works against specificity on a species-by-species basis, the forest type description is more comprehensive in addressing the many parts of the forest ecology. This coarser-scale description has also lent itself to remote data collection, by airplane-based mapping and aerial photography and now by satellite-based imaging.

At coarse scales the shape and composition of individual trees on the ground plane are not seen as discrete objects but as textures and colour patterns. The widely successful use of satellite-based scanners and imaging devices illustrates that we can tell much about the composition and condition of forest cover by reference to signals about colour, reflectivity and penetrability, which tell us little about the numbers and locations of the individual trees but much about them as a forest. Orland, in 1991, described an image-processing approach to represent the systematic colour changes occurring in forest canopy affected by forest insect damage. Filter values were identified that matched different stages of infestation and were applied to the areas of images predicted by forest entomologists to be most susceptible to change. That approach was extended by the development of visual patterns that could be inserted into images to represent large-scale change such as the effects of fire. Others have achieved outstanding representations of forest change at this larger scale using tools such as World Construction Set and its successors, notably the research group at University of British Columbia (see 'Planning, communicating, designing and decision making for large scale landscapes', p. 120). The close connection between the data gathering and management characteristics of remote sensing and Geographic Information Systems at this scale lends itself to the ready representation of newly emerging data or the mathematically modelled results of change such as forest growth, fire spread and pest outbreaks.

The middle-ground challenge

Between these two ends of the scale spectrum lies a region where trees and shrubs are recognizable as individuals of different species, are clumped or dispersed with respect to one another, yet the information gathered and projected about their growth and change is based on sampled data and statistically summarized. In foresters' terms the 'stand' or 'block' is a fundamental unit of forest management, each being defined as an area of relatively homogeneous forest of consistent topographical characteristics such as slope and aspect. While derived from management of natural and plantation forests, the same essential conditions apply to urban and recreational forest, where management actions are taken on individual and recognizable trees, yet information about the forest is maintained as numbers of trees per acre, with little spatially explicit information.

These two different views of the forest create a dichotomy where the representational tools, in seeking to create more realistic images, obscure the fact that the data on which the images are based are summary data – and that essential spatial information has been discarded in the sampling and summarizing process. However, the process of creating a visualization demands that spatial data be re-assigned to the elements of the image. In the absence of site-specific data, some form of random distribution generally accomplishes that. Some developers have created tools that provide visual representations of forest inventory data, such as the Stand Visualization System (SVS) and Envision, a landscape-scale visualization tool, both developed by Robert McGaughey and his colleagues (2003), and SmartForest, developed by the author and his collaborators (Orland 2003). Both take data from US Forest Service inventory and use them to create visual representations of forest stands. In the former case this is done at the scale of a one to four acre (0.45–1.8 ha) plot, in the latter at landscape scale in the context of other stands and including the representation of topography. Both sets of developers have concluded that the realism of the trees represented is beneficial to the credibility of the representation as an image of forest and have striven to provide realistic tree symbols. However, in each case the symbols tend to represent free-standing trees and not the idiosyncratic geometries of trees grown among a possibly heterogeneous mix of neighbours of varying species and densities. Even more importantly, they are poor at representing the ground and middle-layer growth of shrubs and young trees that, while missing from an inventory of trees, can be a dominating factor in determining the visual openness or congestion of the forest seen at the level of a ground-based observer.

Case study – The Gunflint Trail

A recent implementation of SmartForest to represent the outcomes of different forest management scenarios has offered a new perspective on the intertwined issues of representational validity and visual realism. During the July 4 Independence Day holiday in 1999 a powerful windstorm in the Boundary Waters Canoe Area Wilderness resulted in widespread forest blowdowns – areas of completely uprooted or snapped-off conifer and deciduous trees (USDA Forest Service 2000). The area is in northern Minnesota, and across the Canadian border in western Ontario. Studies have been directed at trying to understand residents' and visitors' preferences in relation to the visual effects of the alternative, and controversial, policies that seek to restore the forest cover in some areas and achieve reductions in fire hazard in both impacted and untouched areas (Daniel *et al.* 2003). The survey format used visual representations of the anticipated conditions.

To establish the validity of the visual condition projections, forest data was collected as part of an intensive inventory for ecological modelling purposes (Gilmore *et al.* 2003). The data included the species and size of each stem over 6 mm diameter. The Forest Vegetation Simulator (FVS) (Teck *et al.* 1996) was used to project future forest conditions. FVS is a standard in use across the USA, with extensions tailored to the species and environmental conditions of numerous regions, so that it was possible to represent the specific parameters, time-steps and specifications for a range of forest operations including thinning and planting, the major treatments anticipated for the study sites. The growth model takes into account over- and under-performing trees so that each original tree may be represented by three, nine or more surrogate trees in successive simulation cycles. The model and its many extensions represent mortality among out-competed or senescent trees, and include natural regeneration of both commercial and non-commercial species. Thus, the investigators created projections of forest conditions, based on ground-sampled data, for ten-year time steps to 100 years into the future.

Serious problems arose from the process of assigning spatial locations to the tree records, essential to the visualization process. In depicting single-year representations of forest data, with tools such as SVS, Envision and SmartForest, the summary data from inventory or model can be distributed randomly in a three-dimensional space. However, for such an approach, each new visual representation will invoke a new random distribution of the data. A major challenge that arises in trying to represent multiple time-steps is that of ensuring that trees remain and grow in the location assigned to them in the first time-step. Figure 6.2 indicates the visual differences between three randomly generated views of the same original data.

It is possible to track individual tree records and assign them to the same spatial coordinate in each subsequent time-step if the number of

Applications

6.2 2002 dataset, three different views

trees remains constant, as the distribution sequence generated by the random number generator will be the same as long as the seed value is the same. Unfortunately for visualization ease, the numbers of trees does not remain constant. They die and are replaced, new trees are constantly appearing as seeds germinate and grow, and forest operations may harvest or otherwise remove trees from the tree list. In any of these instances the distribution sequence will be disrupted.

To address this problem SmartForest pre-processes a complete simulation run and inserts 'dummy' records to compensate for changes in the numbers of records – at the beginning of the sequence to represent trees that eventually emerge as regeneration or as under- or over-performing

6.3 2022 (top) and 2052 (below) visualizations of same data set as Figure 6.2

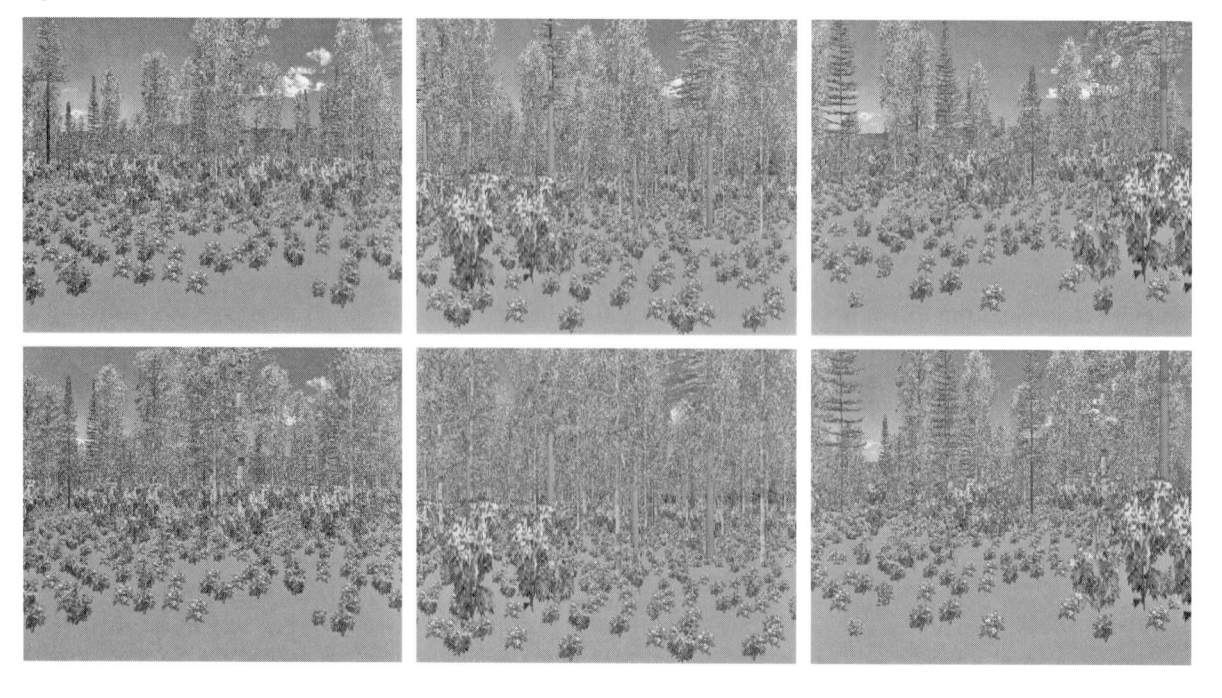

Applications in the forest landscape

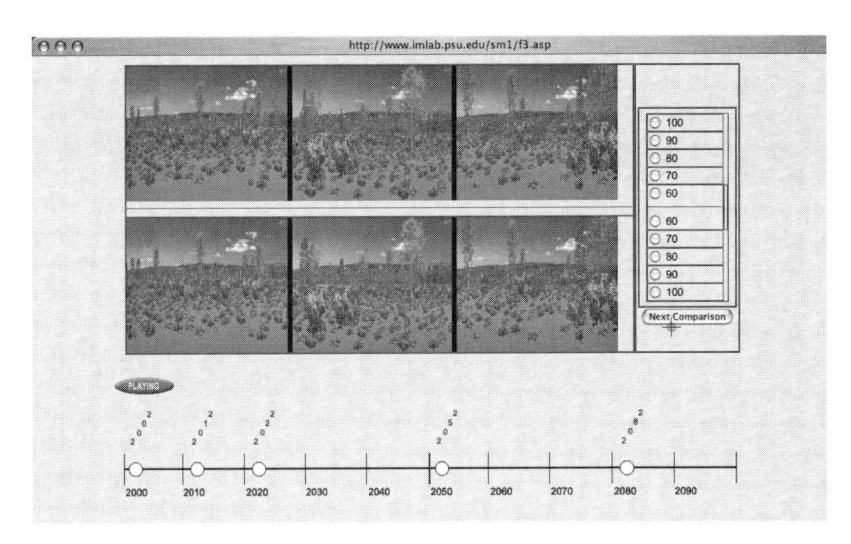

6.4 Typical survey page (image reproduced in the colour plate section)

trees, and at the end of the sequence to represent now dead or harvested trees. Figure 6.3 shows 2022 and 2052 representations of the same dataset as that in Figure 6.2. Looking at all three time-steps together, the left and right panels are most clear in showing the repeat of significant trees and the changes around them. The centre panel illustrates the instance where a central tree, present in 2002 and 2022, has fallen by 2052.

The resultant images were incorporated into an on-line survey instrument (Figure 6.4) that was used to solicit public input in more than 200 face-to-face interviews. Image sets were shown as animations stepping viewers through five time-steps as shown. Each sequence of images was preceded by images of the real setting so that the validity of the base representation could be established, then viewers were asked to focus on the change in forest conditions as represented by the changing images. The viewer responses were remarkable for their degree of acceptance for these admittedly abstract images. The study design asked people to compare the acceptability of different forest management policies shown by the image sequences. Their responses were consistent with our expectations of the likely impacts of thinning and other management practices, suggesting that these calibrated images do meet the need for visualizations of this middle-ground of forest representations.

Conclusion

This section describes evolution in the representation of future landscapes and the unique difficulties represented by the middle-ground. Even given the immense effort described above to achieve a replicable but changing landscape over time, it is clear that much still needs to be done.

Applications

Two major issues arise that perhaps represent core questions for those involved in visualization. First is the necessity at the heart of any visualization to identify a spatial location for each object to be shown. While the complexity and scale of landscapes in this middle range, and the need to anticipate the locations of newly emerged trees, mitigate against achieving complete spatial data, every effort should be made to ensure that the landscape 'behaves' plausibly. Comparison between time-steps appears to require that the same viewpoint be represented at each step, so that growing trees remain in the same place through time and new ones clump or scatter as in nature.

The second issue is to consider whether the object of the visualization might be masked, or should be masked, by other scene artifacts. In the instance illustrated in Figures 6.2 and 6.3, base data was available for those shrubs and forbs present in 2001 but growth and development data for ground-cover and shrub species was not. Although the resulting images of those components were thus not accurate to the anticipated conditions, if such detailed information had been available the improved validity of that aspect of the visualization might well mask changes in the major vegetative component – the trees. Ground-level conditions at the site represented in those images in 2003 included five to six foot aspen seedlings – locations photographed two years previously had rapidly become visually impenetrable tangles of foreground foliage (Figure 6.5).

The concept of calibration is central to ensuring that images are useful to decision making or other judgments, but the experiences reported here indicate that there are critical questions to be addressed before use. In each example above, the trees rest in a matrix of other elements – geological, man-made, and natural – that are each changing at the same time as the object of the visualization. In most cases information will be incomplete and models will be poorly calibrated. In any case, our ability to predict the ephemeral impacts of weather, disease or

6.5 Site conditions photographed in 2001 and 2003

fire is poor enough to render impossible any notion of accuracy in visualization of future events.

Nevertheless, in each of the cases described here the central issue was to represent the impact of change in just one element of a scene – the addition or subtraction of a street tree; the impact of insect damage on forest canopy; and the impact of tree growth and change in a forest landscape. Calibration exists to fit visual images to systematically altered representations of the future and our abilities to do so have improved markedly and continue to improve. If it is possible to limit the role of visualization to just that, then our major concern is to address the validity of the visual and spatial representation of the changing element of the landscape, and in that area good progress is being made.

STUDYING THE ACCEPTABILITY OF FOREST MANAGEMENT PRACTICES USING VISUAL SIMULATION OF FOREST REGROWTH

Ian D. Bishop, Rebecca Ford, Daniel Loiterton and Kathryn Williams

Introduction

Forest harvesting is a controversial topic in Australia. The Australian public has a special affection for the native Eucalpytus forest and the animals it supports. The practice of clearfelling, followed by high temperature burning and aerial sowing of eucalypt seeds (clearfell, burn and sow – CBS) is the most commonly used approach for wood production in tall wet eucalypt forest in south-east Australia (Florence 1996). This creates a short-term scene of apparent devastation within a harvest block – locally called a coupe – which is seen by some members of the public as a ruthless approach, insensitive to the values of sustainability, wildlife habitat and aesthetics. The public and private forest management agencies, on the other hand, see the practice as both the most efficient – in terms of cost of timber removal – and also among the safest and most environmentally appropriate practices.

Forestry Tasmania has established a silvicultural systems trial (Hickey *et al.* 2001) within the Warra Long Term Ecological Research (LTER) site to undertake scientific research into the consequences of alternative forest management practices. The site is an area of wet eucalypt forest with a very dense understorey. The dominant species, and main timber tree, is the stringybark (*Eucalyptus obliqua*). The major understorey species are dogwood (*Pomaderris apetala*), myrtle (*Nothofagus cunninghamii*) and silver wattle (*Acacia dealbata*). Several other species occur in smaller numbers but are economically important: e.g. leatherwood (*Eucryphia lucida*) for bee keepers, celery top pine (*Phyllocladus aspleniifolius*) for boat builders and joiners.

From 1998 to 2003, sections of forest have been harvested to proscribed patterns and the distribution of seed fall, germination and regrowth monitored. Among the harvest and regeneration treatments being rigorously assessed are the following.

- Clearfell, burn and sow: the area cleared is typically about 60 ha, the burn is hot.
- Dispersed retention: a percentage of individual eucalypt trees are retained for a full rotation for fauna habitat and natural seed supply. The slash is partially-cleared using a low-intensity burn.
- Aggregated retention: islands of undisturbed forest are retained for a full rotation for habitat, seed supply (all species) and aesthetics. A low-intensity burn is used.

In order to present the public with an understanding of the full harvest and management sequence – as distinct from the emotive view of a burnt scar – we have created animation sequences covering 200 years of forest life. This is typically two harvest cycles.

Preliminary simulation and assessment

In order to determine which elements of the forest environment are the most important to simulate, and which features of these elements require accurate portrayal, we created an initial set of still images. These were done quickly using simplified versions of the procedures described below. The forest elements included in these initial simulations were based on site observation of and advice from professional foresters. These preliminary simulations were shown to 18 people recruited from organizations with a range of interests in forest management (forest industry, minor species timber users, conservation groups and people from organizations with no formal position on forest management).

Before visiting field sites, participants rated the acceptability of the simulated forest management systems. At each of the corresponding field sites, participants completed a brief questionnaire. They rated the acceptability of the harvesting system used at the site and listed their reasons. Participants then rated the accuracy of the simulation and described any differences between simulation and site that were relevant to their acceptability judgment. The trial simulations did quite poorly on the level of accuracy or realism and there was relatively low correlation between the acceptability judgments made from the simulations and the field sites. Descriptions of differences between the simulations and the field sites highlighted areas for improvement. From this we identified a number of necessary improvements in the simulation process:

- careful rechecking of density and height estimates used;
- less uniformity in the regrowth pattern;
- use of textures better representing the species under regrowth conditions;
- more understorey plants;
- more stumps and logs and other harvest waste – blackened as appropriate;
- bare areas along snig (tree removal) tracks.

Realistic simulation process

Based on our initial experiences and the on-site assessments, we began a new round of simulation development. A key element in the process

Applications

was the selection of the rendering software. Several packages, which could potentially assist with the model development and rendering process, were available. These included 3Dstudio Max, Bryce and the public domain programs Forester and VTP. None of these, however, gave us the level of control we needed to generate a time series of images in which:

- trees were randomly located but remained in the same location at the next time step;
- several different textures were available for each species and were randomly allocated to individuals of that species;
- the textures could be changed at a specific age corresponding to a change in the growth habit of the species;
- the crossed-planes upon which the textures were pasted were randomly rotated – but retained that rotation as they grew;
- within each age class a level of height variation was randomly distributed;
- we could reduce the tree density further from the viewpoint to contain rendering times.

To achieve complete control over the rendered model we chose to use the public domain renderer POV-Ray. This is wholly controlled by text files which describe the scene in the POV-Ray modelling language. We then developed a Visual Basic program to generate the text files.

Determination of forest species mix

Existing data from previous vegetation surveys were used to generate tables of likely species density at selected ages (1, 3 10, 25 and 89 years). Growth formulae have also been developed for the most common species and used to compute the heights of the plants at the key ages (Table 6.1).

Table 6.1 Height and density estimates at different ages for the clearfell, burn and sow condition. Similar tables are used for the other harvest systems

Silvicultural system	Age	Eucalypt Height (m)	Eucalypt Density (stms/ha)	Acacia Height (m)	Acacia Density (stms/ha)	Myrtle Height (m)	Myrtle Density (stms/ha)
CBS	3	4.5	3000	4.3	10 000	0.3	100
	10	12.1	1000	10.3	8000	1	100
	25	23.5	500	14.1	3000	2.5	100
	89	44.1	250	26.5	500	9	200

Distribution of forest mix over terrain

The process of allocating individuals of each species to a location in the terrain began with the creation, using a VB program, of a very large number of random points at coordinates within the minimum bounding rectangle of the chosen coupe. These points were imported, as a text file, into ArcView. For each harvest system, areas of retained trees were defined by Forestry Tasmania officers. These boundaries were digitized into ArcView and then used to cull the random points to leave the new growth points.

Development of individual trees

Real-time rendering performance was not sought, and use of full three-dimensional models of each tree was an option. However, we decided to use a texture mapping approach because:

- the species found in the Tasmania wet forest are very much under-represented in the public domain or commercial three-dimensional tree model libraries, and we had little experience of developing such models; and
- the presence of hundreds of thousands of trees with each modelled as several hundred polygons would have made rendering too slow.

The first step in this process was to visit the Tasmanian forests and take a large number of high-resolution digital photos of each required species at varying ages. For the best results, a uniform background is required so that the green leaves of the tree can be easily distinguished. For larger trees this involved finding examples with clear sky behind them (and ideally the sun directly behind the camera). For smaller plants we either cut the stem off at the base and held it up to the sky, or held a blue tarpaulin behind it. The digital photographs were edited in Photoshop to separate them from their background. Whether this is done by colour selection or outlining procedures, there are always some background pixels that remain, giving the tree an undesirable blue or white outline. However, by adjusting the hue and saturation of all blue and cyan pixels until they matched the colour of the tree, we were able to combat this effect.

Once all the textures of a given species were completed, the red, blue and green brightness/contrast levels of each image were adjusted until the colour-change between the different-aged trees appeared to be smooth but still photo-realistic. The change in shape was also taken into account, with various branches and clumps of leaves deleted, copied and pasted in different places until the trunk-to-tree height ratio and the general branch formations seemed to change gradually over time. The Eucalypt example is shown in Figure 6.6.

Applications

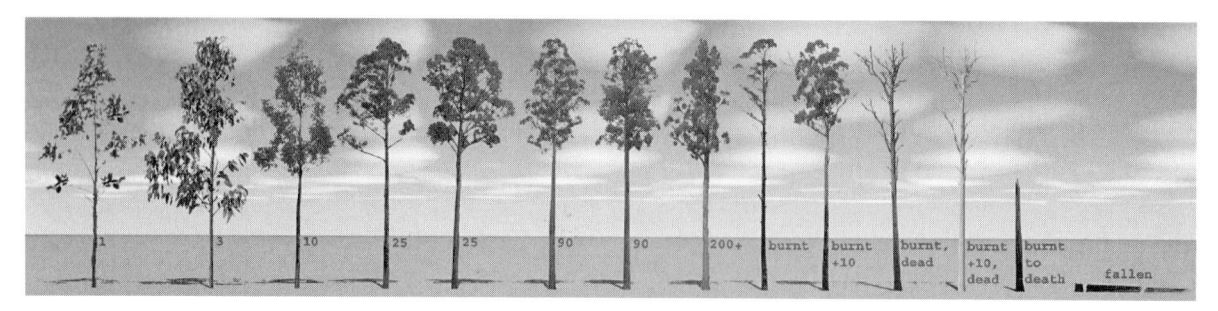

6.6 The textures used for different ages of *Eucalyptus obliqua*, including the effects of wild-fire. Textures were initially created at a common size, then scaled for rendering

Development of scene description file and rendering

In order to render the desired scenes in POV-Ray, three different files were used. The first file was a pre-written POV-Ray '.inc' (include) file, which contained macros describing how each individual tree would behave, the second was the actual '.pov' file (POV-Ray scene description file), which was generated by our VB program based on a range of user-defined factors, and the third was a POV-Ray '.ini' (initialization) file, which specified options relating to the image output.

The main .pov file is written by a Visual Basic application. The user can select from numerous options relating to the silvicultural system, viewpoint, time frame, and vegetation types to be used. This program takes all of these variables, as well as information from files relating to species density over time and random position data, and then writes out the POV-Ray scene description file. This file renders the world in which the trees grow (i.e. ground, sky, lighting, etc.), and calls the different tree macros, passing them positions (vertical position is determined by POV-Ray itself using the defined ground plane), birth dates, death dates, etc., for each tree in the scene. Minor random height and rotation variables are also introduced. The VB application also gives the user options relating to the final image, such as image resolution, anti-aliasing and the number of frames rendered. When the main .pov

6.7 Simulations of harvest systems showing the improvement in realism between the initial and final approaches

file is written, the POV-Ray .ini file is also updated to include these user-specified options.

Due to the massive number of trees in each scene, the render times for each animation were quite substantial. The time a ray-tracer takes to render depends on how many pixels are in the image and the complexity of the calculations it must perform per pixel. The single factor which seemed to most affect our rendering was the amount of transparent space in the tree textures. POV-Ray must ray-trace through each texture until it finds a non-transparent object. Nevertheless, we felt that the time taken was justified by the results. Figure 6.7 shows the progress made as a consequence of the initial public survey.

Animation and display

The frames of the animation were rendered at 3072 × 768 pixels for display via a Matrox Parhelia graphics card. This card can output seamlessly to three data projectors each set to 1024 × 768 resolution. In the final environmental for public consultation, a 6 × 1.5 m screen was used. This was divided into three sections with the end sections turned in by 30° and the images projected from the rear. The animations were developed using two images per year between harvest and age 17 and one image per year for the remainder of the growth phase. The second set is duplicated so that the animation is two frames per year all the way through. The individual rendered frames were com-

6.8 Frames from the animation illustrating the dispersed retention harvest system: (top) immediately post-harvest; (middle) two years later; and (bottom) seven years later (images reproduced in the colour plate section)

bined into a single .avi file for each harvest system and written to DVD. A playback rate of 4 frames per second gives a total run time of 1 minute 40 seconds. Figure 6.8 shows three stages in the sequence of regrowth after a dispersed retention harvest.

Public response

A second validity test was carried out on the refined simulations, using a method very similar to the first test. Nineteen people were recruited for this study. Half were recruited through conservation and timber industry organizations. The other half were not affiliated with any organization with a formal position on forest management.

The study aimed to test both the refined simulations and the proposed methods of presenting them on the large screen. Participants were first shown eight still simulated pictures on the large triple screen and asked to rate the acceptability of the management systems represented. Participants were then shown a time sequence (since the full animations were not yet rendered) for each management system (at years 0, 1, 2, 3, 7, 25 and 88) and asked to rate the acceptability of the full sequence.

The study then visited eight field sites corresponding to the still simulations. At each site participants rated the acceptability of the management system at the site and noted the reasons for their judgments. They then described differences between the site and the simulation that were relevant to their acceptability judgments and rated the accuracy of the simulation. To help them in doing this they were provided with a hard copy of the simulation at A4 size and were asked to use this to remind them how the picture had looked on the large screen.

Six of the eight simulations elicited very similar acceptability judgments to the corresponding field sites. Accuracy ratings were much higher than those given in the earlier study. Participants' qualitative responses identified ways in which the simulations could be improved.

Across all of the harvest systems and most of the age classes tested, participants commented that there was more understorey and ground cover vegetation in the field than in the simulated pictures. In itself, this did not appear to influence people's judgments, but was important in terms of their ability to recognize the forest type as wet Eucalypt forest, which is characterized by dense understorey. In the early years after harvest, participants commented that there was more debris in the field than in the simulations. In some cases this appeared to be influencing acceptability judgments.

Conclusion

The field evaluation was critical to the process of developing simulations useful to the task of assessing acceptability of forest management systems. The preliminary assessments were exploratory, incorporating both quantitative and qualitative responses. They pointed to specific aspects of the simulations that required improvement in the context for which they were developed. Comments on the simulations were made after participants had judged the acceptability of the harvesting system at each field site, and explained their reasons for these judgments. Comments on simulation accuracy were therefore closely related to the criteria used in these judgments. People with diverse perspectives on forest management undertook the assessments. This diversity was reflected in their responses. Participants with an interest in the management of special timber species (for example for boat building) were more likely to comment on the absence of these species. The development of realistic simulations is constrained by resource limitations including computing capacity. Exploratory evaluations should be conducted to ensure that the environmental characteristics that are included are those most relevant to the purpose of the simulation. The results of the second study are particularly interesting in this respect since the level of understorey and debris was criticized in the simulations and yet the acceptability judgments based on the simulations still appeared to be valid. A diversity of views and knowledge among the evaluators helps to ensure that the information content of the consequent simulations meets all needs.

The use of public domain rendering software, in conjunction with our own image specification software, proved to be highly effective in this circumstance and gave us considerable control over simulation parameters. Commercial landscape simulation products (such as VNS) that provide for full georeferencing of terrain and land cover may well produce comparable results.

Acknowledgements

The project was supported by an Australian Research Council Linkage grant, Forestry Tasmania and the Bureau of Rural Sciences. John Hickey provided the essential species height and density data. Gill Fasken produced the first generation simulations.

PLANNING, COMMUNICATING, DESIGNING AND DECISION MAKING FOR LARGE SCALE LANDSCAPES

Jon Salter, Stephen R.J. Sheppard, Duncan Cavens and Michael Meitner

The nature of forest resource management and related efforts in decision support and community awareness demand particular attributes of landscape visualization systems (Orland and Uusitalo 2001). These include: an ability to deal with extensive land areas with highly specialized data sets; an ability to mesh with ecological and economic modelling systems; an ability to represent complex forms (trees and other vegetation) with various levels of detail and realism; a need to convey complex landscape change over long periods of time; and an ability to support the forest managers' strategic, tactical and operational decisions, while explaining the complexities of forestry to an increasingly interested and skeptical lay-public at local and global levels. Not surprisingly, the application of visualization to forestry is itself multi-facetted, and changing rapidly as new technologies, new public expectations and new corporate/NGO policies emerge.

The complicated systems needed to model and represent digitally the complexities of the forest landscape in various presentation modes, translate to a relatively small number of research facilities around the world geared at developing, testing and applying visualization systems to a range of forestry applications.

One of the centres for research in forest landscape visualization is the Collaborative for Advanced Landscape Planning (CALP) based at the University of British Columbia (UBC) in Vancouver, Canada. The composition of the team at CALP reflects the widening scope and system requirements that characterize the rapidly evolving interdisciplinary field that landscape visualization represents: members include researchers from forestry, landscape architecture, planning, environmental psychology and computer science. CALP's goals are to develop better ways of planning, communicating, designing and decision making for large scale landscapes.

Within this context, CALP specializes in the development, use and evaluation of visualization tools and procedures. Visualization related projects include:

- development of a real-time renderer of forested landscapes (Cavens 2002);
- development of new interaction techniques for landscape visualization using tools such as laser-pointers (Cavens and Sheppard 2003) and other information/visualization interfaces (Sheppard and Salter 2004);

- integration of forest harvesting, growth and yield and species habitat models with visualization techniques (e.g. Meitner *et al.* in press);
- evaluation of the benefits of immersive displays;
- use of visualization in public planning and decision-making processes for forestry (Sheppard 2000);
- use of visualization in First Nation's consultation processes (Lewis 2000); and
- assessment of the representational validity and ethical considerations of different visualization techniques (Daniel and Meitner 2001).

This chapter highlights progress made to date with the use of visualization in one example drawn from these projects, and summarizes some key research findings and implications.

The Arrow Innovative Forest Practices Agreement (IFPA) Project

One of the recent projects that CALP has been involved with is the Arrow Innovative Forest Practices Agreement (IFPA) project. This project was funded by the provincial government as one in a series of initiatives aimed at investigating innovative methods of conducting forestry in the province of British Columbia.

CALP was involved in this project as part of an interdisciplinary team from the University of British Columbia that included researchers from ecology, timber supply analysis, wildlife biology, hydrology, recreation and the social sciences. This interdisciplinary group was charged with looking at ways of conducting sustainable forestry in the Arrow Timber Supply Area, a 754 000 ha area in the south-eastern portion of British Columbia. Multiple forest management scenarios were devised by the interdisciplinary UBC team, in order to examine the effects of alternative techniques on the sustainability of forestry in the Arrow TSA. CALP's role in this process involved both technology development to integrate visualization with models being used by the UBC team, and public consultation to determine the social priorities for, and acceptability of, sustainable forest management in the region. This involved both the development of new visualization software and the application and testing of visualizations in a new public planning process (Sheppard 2003).

Technology development

In order to analyse the management scenarios proposed for this project, it was necessary to incorporate several disparate modelling packages into a cohesive scenario analysis tool. While several of these packages

Applications

6.9 A schematic representation of the landscape visualization system as applied to the Arrow IFPA study

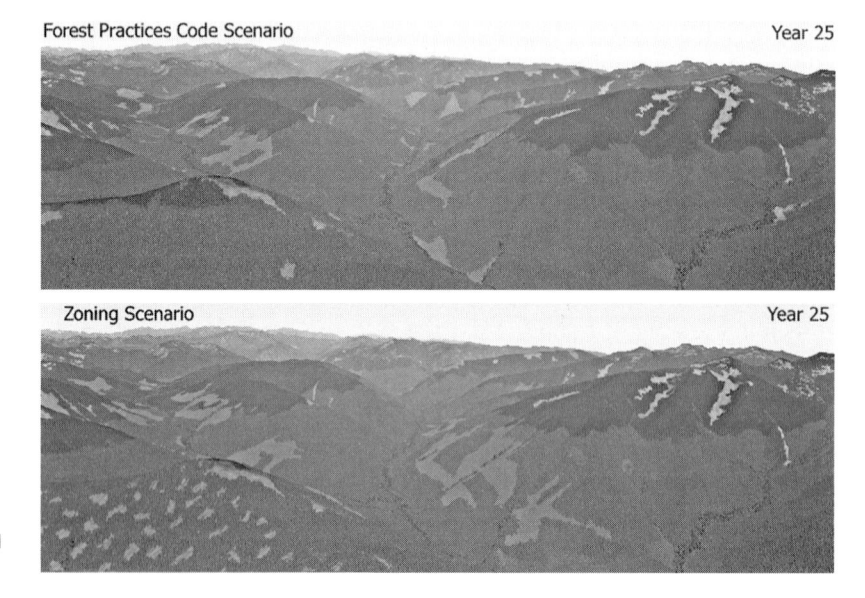

6.10 Visualization of alternative forest management scenarios at Year 25, showing overlays of habitat modelling for one bird species (images reproduced in the colour plate section)

have previously been integrated, they had never been combined with a visualization tool. In 2001, CALP researchers created a system for combining data inputs from the FORECAST forest stand attribute model (Kimmins et al. 1999), the ATLAS landscape level forest harvesting model (Nelson 2003), and the SIMFOR species habitat model (Wells and Bunnell 2001), to create model-driven simulations of different forest management scenarios over time, using World Construction Set as the rendering engine (see Figure 6.9 for a schematic representation of the model integration system).

The CALP visualization system was designed with flexibility in mind, so that it is not limited to the modelled inputs used in this study, nor to the use of World Construction Set as a renderer of the generated visualization information. An example of the visualizations generated using the system can be seen in Figure 6.10.

Application

The Arrow IFPA project included a significant consideration of the social components of forested landscapes. Some of the social inputs included a region-wide socio-economic survey and stakeholder analysis and focus group meetings within the context of a Multi-Criteria Analysis (MCA) for a landscape unit (41 000 ha) within the Arrow Timber Supply Area (Sheppard and Meitner 2003). The purpose of these public involvement mechanisms was to help develop criteria and indicators for Sustainable Forest Management (SFM), assess local public priorities for landscape management, and evaluate alternative forest management scenarios (see Figure 6.10). The visualizations created for this project were used in the focus group MCA meetings at different stages during the overall process, to help explain the effects of the different forest management scenarios over time. They were presented in split-screen (multiple scenario) formats via projection screen, in 'time-lapse' sequences of ten-year increments from Year 0 up to almost 200 years, from both an oblique aerial view and a ground-based viewpoint. The visualizations were used as one part of an information package that included background information, resource mapping, expert opinions, and draft criteria and indicators for sustainable forestry. The focus group participants were asked, among other things, to give their direct overall preferences for the scenarios, based in part on what they had seen in the visualizations. This allowed comparison of direct preferences for scenarios against expert evaluations weighted by previously obtained stakeholder group priorities.

As part of the focus group exercise, the stakeholders were also asked to rate the effectiveness of the tools provided to them in order to assist their deliberations in the overall landscape planning process. While data analysis is not yet complete, preliminary results indicate that use

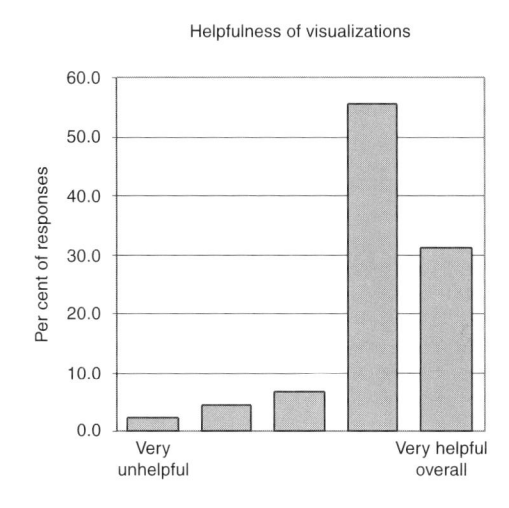

6.11 Stakeholder evaluations of the helpfulness of landscape visualizations used in the public planning process in the Arrow IFPA project

of the visualizations ranked second only to the presence of a neutral facilitator in terms of their usefulness to the process, ahead of maps, criteria weightings and expert evaluations. Overall, most participants found the use of visualizations helpful or very helpful in their deliberations (Figure 6.11). However, some stakeholders did register concerns over particular aspects of forest representation in the imagery, such as providing too rapid an impression of restoration of mature forest characteristics after harvesting (Meitner *et al.* in press).

Implications

The finding that fairly realistic visualizations of planning options are seen to be useful by lay-communities has been encountered elsewhere, ranging from urban communities (e.g. Al-Kodmany 1999) to rural aboriginal communities (Sheppard and Lewis 2002). At CALP, we have heard repeatedly from communities that visualizations can be a major help in understanding forestry issues and providing an avenue for dialogue. Lewis (2000) has demonstrated how photo-realistic visualizations substantially increase a community's ability to articulate its preferences for the landscape and provide a more meaningful type of input to forest management plans (Sheppard *et al.* 2002). The usefulness of such images to forest managers and planners themselves has also been observed, in identifying data or modelling flaws and raising questions about the scenarios emerging from expert methods (Meitner *et al.* in press).

However, the full extent of the influence of visualizations on the forest planning and decision-making process has not been mapped out, and skeptics from both the public and from natural resource disciplines raise important issues of what is true and what is misleading (McQuillan 1998; see also Chapter 5). Much more research and testing of these issues as applied to forestry is required before we can strengthen our guidelines for the use of landscape visualization in forestry.

THE ROLE OF LANDSCAPE SIMULATORS IN FORESTRY: A FINNISH PERSPECTIVE

Liisa Tyrväinen and Jori Uusitalo

Possibilities of using visualization in Finnish forestry

At its best, a forestry visualization tool embodies several design goals which may vary according to the particular user. Forest owners can use forest visualization in demarcating appropriate areas for logging. A forest owner may not be interested in knowing in detail the different characteristics of the trees but is certainly attracted by comprehending the commercial value of their forest holding or logging area of interest. Comparison of the commercial value of logging areas need not be separated from other valuations. The current values of each commercial wood assortment may be linked to the visualization system's database after which the value of each tree, tree group or stand may be queried by a simple mouse click or highlighted with different user-defined colour codes (Uusitalo *et al.* 1997; Uusitalo and Orland 2001). Visualization enables the forest owner to compare the financial benefits of each area of interest in readily understood form and aids the owner to better contrast monetary benefits with non-monetary ones (Orland *et al.* 2000).

The role of wood procurement managers in industry is to buy stands that meet market needs. Despite thorough annual and monthly planning based on factories' orders, wood procurement is a very dynamic process where the demand of each wood assortment may vary rapidly. Some wood assortments may be extremely desirable at one instant but may be totally rejected at another. Most wood assortments' demand varies by season and economic trend while some high value wood assortments are highly desirable at all times (Uusitalo *et al.* 1997).

Advanced forest visualization tools may possess extremely valuable features with the potential to aid wood procurement managers to judge with greater accuracy the distributions of sizes and qualities of the forest resource which are critical to purchase decisions (Uusitalo *et al.* 1997). Extensive use of computer graphics enables the user to visually classify the trees according to different tree characteristics. With the help of special data-selection tools, the user can customize a classification for the characteristics and define a colour palette to represent each class. This colour classification enables the manager to efficiently envision the especially advantageous characteristics of a stand or forest holding (Orland and Uusitalo 2000).

Moreover, forestry in Finland is submitted to scrutiny by different types of governmental and non-governmental consultation organiza-

Applications

tions. Due to an increasing number of urban forest owners living far from their forest resources, local forest owners' consultation associations have gained an important role in the wood trade in Finland. The major tasks of the local manager are to consult forest owners on wise management practices and to control wood trade (Uusitalo *et al.* 1997). Visualization is now seen as a powerful tool to assist ever-increasing numbers of urban forest owners with little technical forestry knowledge in understanding forest dynamics and their huge impact on forest and scenic resources over time (Orland and Uusitalo 2000). In order to show wood buyers what is available from their owners' forest, the local managers need an effective tool to communicate complex multi-dimensional data. Some forest visualization software have an ability to demarcate stand boundaries with different colour codes which can help the manager to separate different stands and forest holdings at one view. This feature has great value in Finnish forestry due to the small average size of forest holdings and stands (Uusitalo and Orland 2001).

During the past decades, due to the structural changes in agriculture and forestry, the countryside has changed from being a place of primary production and is becoming a place of recreation and tourism services production. Also, the motives of private forest owners in Finland have changed towards non-consumptive uses. Today, one-fifth of private forest owners consider scenic and recreational values as the most important management objective, and half of owners consider them to be as important as income from wood production (Karppinen *et al.* 2002). Moreover, urbanization has challenged urban woodland planners and managers to create and maintain attractive environments to meet the wide array of demands from urban people. In the above-mentioned forest areas, a balance between traditional economic and the less tangible amenity benefits of forests has to be achieved. In this context, evaluation of visual impacts of forest management practices is crucial in order to meet the expectations of tourists, recreationists and other users (Nousiainen *et al.* 1998; Tyrväinen and Tahvanainen 2000).

Visualization tools could serve as facilitators in forest planning and design in areas with scenic values, tourism development areas and in peri-urban and urban forests (Tahvanainen *et al.* 2001; Tyrväinen *et al.* in press). First, the visualization can be applied in landscape preference research to illustrate various management options for different interest groups such as local residents, experts and (other) decision makers. The information relating to the preferences of various groups can be fed into forest planning systems, for example, to simulate alternatives that are socially acceptable for the wider user groups. Second, computer-aided illustrations can be used for presenting and communicating new management ideas and options in planning. The future scenarios of management and development lines of forests could be discussed through the use of visualization (Tyrväinen *et al.* in press). The tool would be helpful in finding a common goal or sharing ideas between professionals and/or wider audiences. Third, the tool can be used in interactive planning sessions to illustrate particularly the visual consequences of management options and to present different development scenarios. These various ways of using visualization will help in gathering, in particular, local information related to a planning area and learning about the stakeholders' opinions, values and preferences related to future development plans of the area.

Review of three forest landscape simulators

So far, two different types of technology have been used for visualizing forest resources in the context of making decisions about forest management. Simplified computer graphic representations are able to depict the presence of plant species, size classes, etc. found in forest inventory databases. These approaches usually lack the ability to represent detailed aspects of forest landscape composition and thus are unable to achieve the visual realism that might be needed for a specific evaluation. In cases needing such improved visual fidelity, calibrated photographic images have proved their competence but for situations demanding strong validation of the visual conditions it is more difficult to demonstrate strong relationships to underlying tree data. These two categories have been called geometric modelling and video imaging, respectively (Orland 1988; McGaughey 1997).

There are three virtual landscape simulators, FORSI, MONSU and SmartForest, which are partly or totally developed in Finland and are thus able to illustrate forest landscapes in Finnish conditions. In Finland, 75 per cent of the land area is covered by forests, and typical views are close-ups with small scale and fine features. All approaches are based on the use of map information, a digital elevation model (DEM), compartment data of the target area, and visual objects.

FORSI, a commercial landscape simulator is intended to fulfil the needs of practical visualization in forestry organizations. The system has been developed by the Finnish private enterprise Instrumentointi Oy. The Forsi-simulator has a high degree of fidelity in rendering forest scenery with a texture-mapping technique. The two-dimensional visual objects represent the main elements of a forest landscape (trees, shrubs, undervegetation, logging residue). The objects are generated from digitized photographs, and therefore the program produces rather photo-realistic images, in particular when describing scenes from a distance. The tree library of the program consists of main tree species photographed in commercial forests in southern Finland (Figure 6.12). Nevertheless, additional tree species and objects, such as houses and recreational facilities, can be added to the library or included in the pictures manually.

Applications

6.12 Example of an illustration of the Forsi-simulator

In FORSI, the average resolution of the tree textures is 72 pixels per inch, which means that the tree symbols are reproduced in the same size as in original photographs. One problem regarding the technical accuracy of the images is that the demands for resolution and colours increase the closer the viewer gets. Today, resolution in FORSI is not yet satisfactory for public consultations about near-distance scenes (Tyrväinen and Tahvanainen 2000).

FORSI includes the possibility for simulating movement in the landscape, but the viewer is not part of the landscape model as in the SmartForest system. The movement is realized through choosing either a vertical or horizontal location for the viewing point manually, similar to MONSU. The latest version also has real-time movement in the landscape. The simulations of individual forest operations such as clear-cuts and thinning can be illustrated by manipulating the compartment data manually. The program is also able to illustrate the effects of summer and winter seasons as well as atmospheric effects. Because FORSI is a commercial product, its price is significantly higher than that of the other two landscape simulators. The strength of FORSI is the flexibility to interact with many other forest planning tools commonly used in Finnish forestry as it possesses sophisticated conversion tools to retrieve GIS-data from various commercial GIS-products and forest inventory data systems.

MONSU, the multiple-use forest management planning system, was developed at the University of Joensuu for the purposes of teaching and forest planning at the farm and regional levels in Finland (Pukkala *et al.* 1995). The illustrations of forest landscapes in the MONSU system are automated computer line graphic drawings based on tree and site parameters included in present forest planning systems. Trees can be illustrated by three different quality levels, and the current technical accuracy of the illustrations is fair. However, the use of the highest quality level slows down the illustration drawing process considerably, particularly in distant scenes. The tree symbols are differently coloured two- or three-dimensional graphic symbols, whose species and size distribution correspond to the local tree populations as described in inventory data.

Applications in the forest landscape

Although the illustrations include some elements of the ground vegetation such as berries and mushrooms, the rest of the ground layer is represented only with different colours. The special features and details of a particular landscape (buildings, shrubs, stones, special shapes of trees and single trees) are absent. Thus, MONSU produces more or less standard landscape pictures, rendering it unsuitable for areas which have, for example, special scenic value. However, the accuracy of the illustrations depends to a large extent on the viewing distance (Tyrväinen and Tahvanainen 1999, Karjalainen and Tyrväinen 2002).

The main advantage of MONSU is that the use of forest inventory data is flexible and efficient, because the system is compatible with available forest inventory or satellite data. The program is also connected to a forest planning system, which means, among other things, that the evaluation of the scenic impacts of alternative forest plans is easy. Simulation of temporal landscape changes by different management regimes is easy both at forest stand and at forest area level. The method enables a flexible assessment of both close-up and long-distance scenes from several viewpoints with updated forest data. Movement in the forest can be simulated by choosing viewpoints along a path and by illustrating the landscape scenes selected, but the capacity of current PCs does not allow real-time movement. However, animations can be prepared by saving illustrations consecutively, for example, from chosen points along a trail and viewing them in sequence.

Moreover, the MONSU program is able to illustrate the effects of seasons (summer, autumn, winter) and atmospheric effects such as fog (Figure 6.13). The program is inexpensive and easy to use on a PC. The recent research project has resulted in a new user-interface of planning software that enables the participation of several evaluators. The latest additions in MONSU visualization are the VRML (Virtual Reality Modelling Language) files that MONSU can generate from a user-specified view. Internet programs like Internet Explorer and Netscape Communicator can interpret these files and these programs in fact generate the visualizations using photographs of trees. A good feature of

6.13 The Monsu-simulator can illustrate the effect of the different seasons on the landscape (images reproduced in the colour plate section)

Applications

the VRML visualization is the possibility of moving smoothly through the forest, which yields a virtual reality effect.

SmartForest, developed at the Imaging Systems Laboratory, University of Illinois in collaboration with the USDA Forest Service and the University of Helsinki, Finland, does possess advanced tools for moving and interacting within and with a forest setting (Orland 1994; Uusitalo and Kivinen 2000). SmartForest comprises two different modes: management mode and landscape mode. Management mode is a simplified presentation of the real forest conditions that enables quick and efficient query and analysis of the various characteristics of forest stands and single trees. The tree trunks are drawn as simple trunk-height bars with a variety of shapes, depending on species and individual tree characteristics (Figure 6.14). In this mode the ground layer is represented by different colours. In the landscape mode, trees and water are represented as texture-mapped objects and the ground is wrapped with realistic two-dimensional ground images generated from digitized photographs. Tree textures have a size of 256 × 256 pixels, and the ground texture has a resolution of 512 × 512 pixels (30 × 30 m grid). Therefore, the quality of illustration in this mode, in particular in the distant scene, is close to photo-realistic. However, in the Finnish version, the photographic database is small at the moment, including only the three main tree species in Finland.

In SmartForest the user may view the ground level, walk between the trees, view large forest areas from user-defined aerial height, and classify stands and trees by highlighting them with different colours. In most cases the user moves the cursor within the view while holding different mouse buttons to achieve longitudinal, rotational and vertical

6.14 The SmartForest Management mode: trees are presented as simplified icons to facilitate quick analyses. Colour classification of the trees enables the manager to efficiently envision the especially advantageous characteristics of a stand or forest holding

movement. The main advantage of the system is that it allows flexible real-time movement in virtual forest landscape which is, however, limited to a straight north–south direction (Orland 1994; Uusitalo *et al.* 1997). To enhance interaction speed, a number of compromises have been made. First, the program displays a reduced number of trees when moving. Another means of enhancing interaction speed is to reduce the 'horizon' in the images and to keep the tree symbols rather simple. This also reduces the ability to represent local scale impacts realistically. The speed of producing illustrations is heavily dependent on the size of the horizon and computer capacity. In the landscape mode a typical case of rendering the forest with a P233-equipped PC varies from a couple of seconds (horizon 3 per cent) to several minutes (horizon 100 per cent) (Uusitalo and Kivinen 2000). The simulations of forest management operations are less easy to conduct than in MONSU. They are realized through a manipulation of the tree data, which increases the costs of producing the illustrations.

Applications of the systems in practice

Few developments in forestry have received such an enormous amount of enthusiasm over recent years as forest visualization. Despite the excitement that it has generated and the market potential predicted for it, the use of the current visualization software in Finland has so far been restricted mainly to research projects and case studies. The case studies have shown that these tools can be used as a research tool and in forest practice. These experiments include comparing usefulness of computer graphic drawings to panoramic photographs in studying public preferences, use of computer graphics drawings in assessment of scenic impacts of farm-scale land-use planning involving local inhabitants, landowners and tourists, and also studying social acceptability of different regeneration methods used in forestry (e.g. Nousiainen *et al.* 1998; Tyrväinen and Tahvanainen 1999).

Despite the recent acceleration of graphic performance capabilities of personal computers, there are no reasons to expect that 'virtual' forest management will rapidly replace existing forest management procedures. One of the biggest obstacles in applying visualization in practice is the lack of appropriate information as well as the labour intensity of combining information from different sources and formats (Uusitalo and Orland 2001). Local forest management plans are today still based on databases comprising the mean values of different tree characteristics. Full utilization of the attributes of a virtual forest would require reliable information on diameter, height and quality distribution of each species (Rautalin *et al.* 2001). Since it is evident that there will be increasing pressures in the future in utilizing various forest visualization applications we will be forced to improve the existing forest inventory procedures in order to benefit from the power of these tools (Orland and Uusitalo 2000).

Applications

In the planning of scenically valuable areas and urban forests the demand for correspondence between the illustration and the details of the real world is even higher than in timber production forests, because the evaluators are generally non-professionals. Important characteristics of a good system in collaborative planning are high interactivity, real-time movement possibility, integration possibility of different sources of spatial data and the ability to illustrate changes in the environment in a realistic manner from both near and distant viewpoints. While many of these properties exist in current landscape simulators they are not all at present incorporated in the same system. For example, forest visualization tools linked to forest planning systems are strong in simulating forest growth and temporal changes of forests, but they have difficulties in illustrating the elements of the built environment. In urban forest planning, however, both of these elements should be illustrated side by side. In contrast, in the different CAD-programs used mainly in park planning systems, the built environment is handled more easily but in these software packages the links to enable the use of spatial forest data or growth models are often missing.

CHAPTER 7

APPLICATIONS IN THE AGRICULTURAL LANDSCAPE

INTRODUCTION
Ian D. Bishop

New techniques for planning and management of the agricultural landscape have not been technology- and visualization-focused in the same way as those for the forest landscape. Changes occur more slowly; ownership patterns are different; there is extensive cultural overlay and complex farmer-support systems. Agricultural landscapes are also very different in different parts of the world depending on climate, technology, culture and government policy. Vos and Meekes (1999: 3) describe the European context and ask the universal question:

> Modern society increasingly utilizes landscape in a great variety of ways and for many purposes. This poses a complex pressure on cultural landscapes, threatening landscape qualities. Therefore planners and managers are facing the question: how can a sustainable future for old cultural landscapes, based on sound economics and the commitment of all actors be achieved?

The case studies in this chapter all begin with the premise that appropriate techniques for landscape design, planning and management include visualization. Yet, although visual classification and preference studies in rural areas have been the basis of academic study for some years (Zube 1973; Orland 1988; Schauman 1988; Gimblett 1990; Cooper and Murray 1992; Lynch and Gimblett 1992; Angileri and Toccolini 1993), visualization is a comparatively recent addition to

Applications

the agricultural landscape toolkit. The slowness to embrace visualization for rural studies seems to be because differences in urban or forest conditions could be illustrated through less realistic visualization tools than changes in the more subtle agricultural landscape.

Even today, much of the change that is illustrated in agricultural landscape in support of planning or management involves particular elements including trees, hedgerows or farm buildings. For example, Gomez-Limon and de Lucio Fernandez (1999) used image manipulation techniques to illustrate different tree densities in the Spanish Dehasa landscape. As stock densities have decreased, tree densities are increasing – but is this what residents and recreators want for this environment? A similar approach is found in Hunziker (1995). In both Spain and Switzerland the existing condition was also the preferred condition. Maintenance of the status quo is also central to work by Hernandez *et al.* (2003) who used image processing to explore the shielding of agro-industrial (farm) buildings using vegetation. They used the same technology to also change the hue, saturation of lightness of farm and village buildings to develop guidelines for new agricultural building developments. Tress and Tress (2003) developed alternative futures for a small Danish town and developed detailed visual representations in oblique aerial view using image manipulation.

All these examples were based on the static image created by image processing. While this can be valuable in certain circumstances the view expressed repeatedly in the studies described in this chapter is that:

- spatial accuracy should be assured by a direct link between GIS and visualization, and
- an interactive system provides much stronger opportunities for effective public consultation than individual images.

Specifically:

- Andrew Lovett's work (see 'Designing, visualizing and evaluating sustainable agricultural landscapes', p. 136) explores options for representation of heritage landscapes with three-dimensional models created directly from two-dimensional mapped information in WWW-compatible VRML format. High levels of realism were not the priority; more important was to give farmers interactive access.
- Christian Stock and Ian D. Bishop (Helping rural communities envision their future, p. 145) describe the development of an interactive 'envisioning' system for use in community workshops. A key element of this work is the interactive linkage of the visualization system to GIS (for orchestrating land use changes and modelling non-visual outcomes) and hand-held computers (for participant input of preferences and votes). Because interactivity is the key design feature of the system, the level of realism remains limited.

- Philip Paar and Jörg Rekittke (*Lenné3D* – walk-through visualization of planned landscapes, p. 152) describe their work on development of a software package in which a high level of realism, combined with interactivity, is the key software design feature. Significant algorithm development is the key, in their view, to making a tool which landscape architects and planners will embrace.

Providing both high realism and interactivity is becoming feasible because of increasingly powerful computers and better algorithms. We need then to consider how best to use this new capacity. What other tools do we need within a visualization-based decision environment?

DESIGNING, VISUALIZING AND EVALUATING SUSTAINABLE AGRICULTURAL LANDSCAPES

Andrew Lovett

Research context and objectives

In England, as in many other parts of Europe, there is currently much debate regarding the future of the countryside, (e.g. Countryside Agency 2003). These deliberations reflect a range of pressures, including economic difficulties in the agricultural sector, concerns regarding habitat loss, problems of social exclusion, and the possible impacts of climate change (DETR/MAFF 2000; Countryside Agency 2002). It is widely argued that greater economic, social and environmental sustainability is desirable, but achieving such an integrated balance is far from straightforward, especially in regions where there are sensitive landscapes and multiple (possibly conflicting) land uses. Programs such as the Land Management Initiatives supported by the Countryside Agency and the Lifescapes project of English Nature are currently investigating the scope for achieving greater rural sustainability at a landscape scale, but it is also becoming apparent that new tools and methods are needed to facilitate the implementation of measures such as Landscape Character Guidelines or Biodiversity Action Plan targets. This is particularly true with respect to means of engaging stakeholders in more participatory and collaborative decision making processes.

A case study that sought to address several of these issues was carried out in 1997–1999 as part of the Global Environmental Change program funded by the Economic & Social Research Council in the UK. The research involved a multidisciplinary team and examined the scope for achieving a more sustainable agricultural landscape through the concept of 'whole landscape management' (Cobb *et al.* 1999; O'Riordan *et al.* 2000). A study area was selected that centred on the Buscot and Coleshill Estate in Oxfordshire, southern England. This estate is owned by the National Trust (a UK heritage charity) and was chosen because it fulfilled a number of criteria (i.e. it represented a distinct landscape unit, included some protected or designated areas, and contained farms with a mix of ownership, specializations and operating constraints). Figure 7.1 shows the main topographic characteristics of the area, including the alluvial valleys of the upper Thames and a tributary, the River Cole, as well as a section of the limestone Midvale Ridge. The total area studied covered 8200 ha, and involved 34 farms, 11 of which were occupied by National Trust tenants. The main elements and objectives of the research were as follows.

- To assemble baseline ecological and land cover information for the area and incorporate this into a GIS database.

Applications in the agricultural landscape

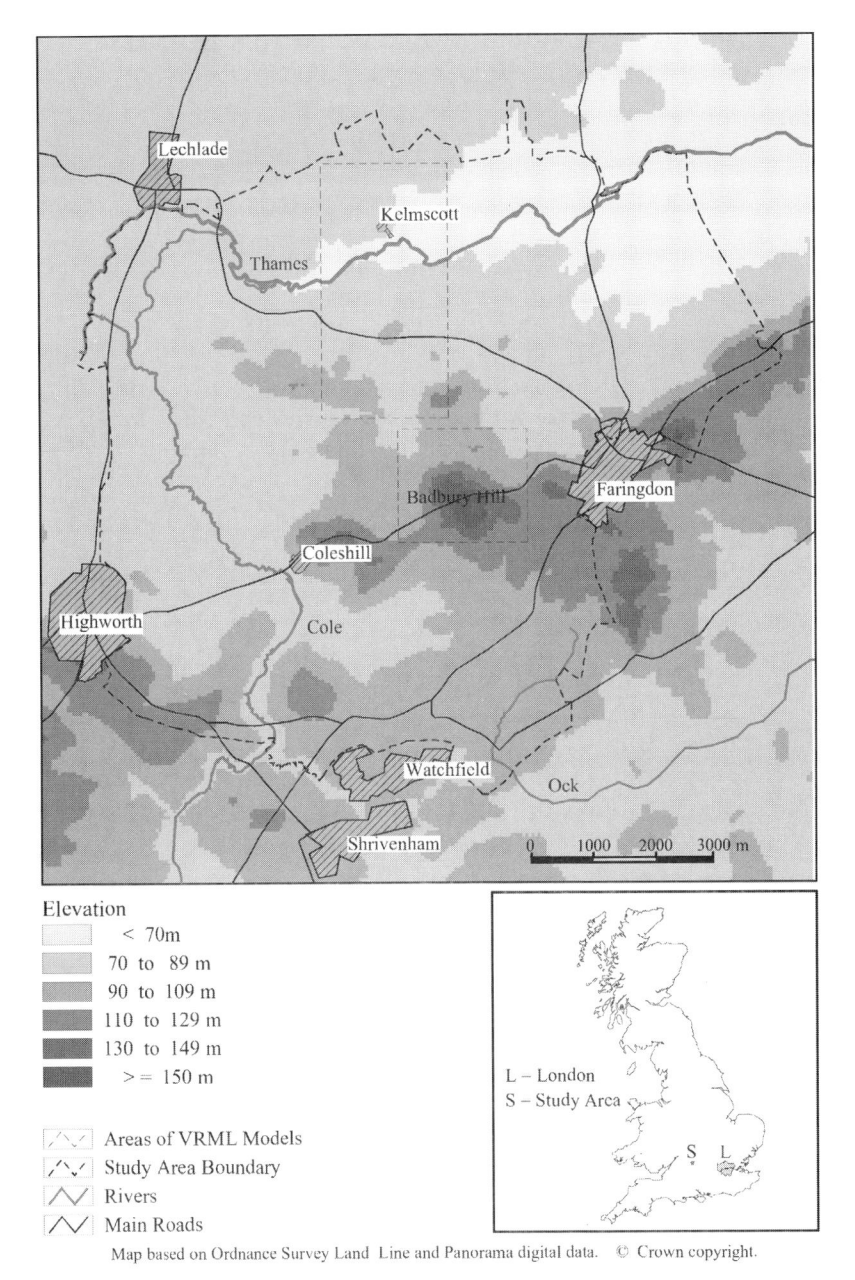

7.1 The location of the study area

- To ask farmers about any changes in management they might undertake in the absence of major policy alterations (giving a 'business as usual' scenario).
- To consult a representative selection of stakeholders and generate three additional scenarios of future whole landscapes that would benefit biodiversity conservation and/or amenity.

- To map the landscape changes under the four scenarios and construct three-dimensional virtual reality models for parts of the study area that would allow viewers to 'walk through' the potential future landscapes and gain a deeper sense of the alterations involved.
- To ask all farmers and managers to respond to each of the scenarios in terms of the circumstances in which they would agree to participate in an agri-environmental plan for whole landscape management.

Initial fieldwork and scenarios

During the first phase of the research an ecological baseline survey was carried out across the study area. This included the collection of information on field boundary and hedgerow characteristics. A programme of structured interviews with farmers was also undertaken. These discussions covered subjects such as the farmers' agricultural practices, financial situation, management intentions and likely responses to possible changes in European Community policy (Dolman *et al.* 2001).

Based on information obtained during the ecological fieldwork and the survey of farmers, as well as discussions with a range of stakeholder organizations (e.g. representatives from local government and environmental NGOs), four scenarios for the future landscape of the study area were devised. These were as follows.

- *Scenario 1 – Business as Usual* was based on each farmer's own plans for future land management. The main source of information was the survey of farmers. Few farmers indicated that they were planning major changes to arable operations, none intended to cease dairying, and only five stated that they would abandon beef production (mainly as a result of difficulties arising from the BSE crisis). As a consequence, the landscape changes in this scenario were rather limited, involving a small shift from arable to grassland and a few habitat protection measures (e.g. buffer strips around streams).
- *Scenario 2 – Landscape Character* focused on maximizing visible amenity. The scenario was based on discussions with stakeholder organizations and information in a variety of planning documents. As biodiversity provides important amenity value, the proposals included some conservation management, but with a focus on popular species of flora and fauna. The main types of landscape changes included hedgerow restoration, conversion of some fields near rivers to grassland, and an increase in deciduous woodland (mainly to screen urban areas or to act as linear features along roads and rivers).

- *Scenario 3a – Biodiversity Conservation* was designed to deliver substantial nature conservation and biodiversity benefits. It was constructed following detailed discussions with statutory and non-statutory organizations (e.g. English Nature and local County Wildlife Trusts) and also reflected policy recommendations for several habitat types. Blanket compliance was assumed across all farms, the main landscape changes being a reversion of floodplain farmland to extensive grass and marshland, increased riparian woodland, hedgerow restoration, the provision of buffer strips around all watercourses, and uncropped margins for arable fields.
- *Scenario 3b – Supplemented Biodiversity Conservation* included all components of Scenario 3a, together with other measures for specific locations in the study area. Examples included the conversion of fields around springs on the Midvale Ridge to rough grassland, and the creation of large scrub or grass buffer zones around designated wildlife sites. The landscape implications of this option were therefore the most substantial of all.

Visualization of the landscape scenarios

Providing visualizations of the possible future landscapes was considered an important means of engaging stakeholders in discussions about the scenarios. The approach adopted was to create a detailed GIS database of the current land use in the study area and then update this to generate separate layers for each of the scenarios. Subsequently, large (A0 size) printed maps of the entire study area were produced for each scenario and three-dimensional landscape models were created for selected smaller zones using Virtual Reality Modelling Language (VRML).

Ordnance Survey (OS) Land-Line vector mapping and Land-Form Panorama digital elevation models (see www.ordnancesurvey.gov.uk) provided the framework of the GIS database. Considerable editing was necessary within the Arc/Info GIS software to convert the raw Land-Line data to a form that was suitable for visualization purposes. This included adding attribute information from paper maps, the ecological survey and farmer interviews to help code different types of polygon or line features (see Lovett *et al.* 2002). Once a database depicting the current situation had been constructed, equivalent map layers for the different scenarios were created using a mixture of manual editing and GIS macros (e.g. to generate buffer strips and field margins in the Biodiversity Conservation scenarios). Figure 7.2 shows an excerpt of the land cover data for part of one of the VRML model areas marked on Figure 7.1. This map depicts Kelmscott village and the nearby River Thames under Scenario 3a (Biodiversity Conservation). Several land cover categories have been amalgamated for purposes of display, but the level of plan detail in the database is readily apparent.

Applications

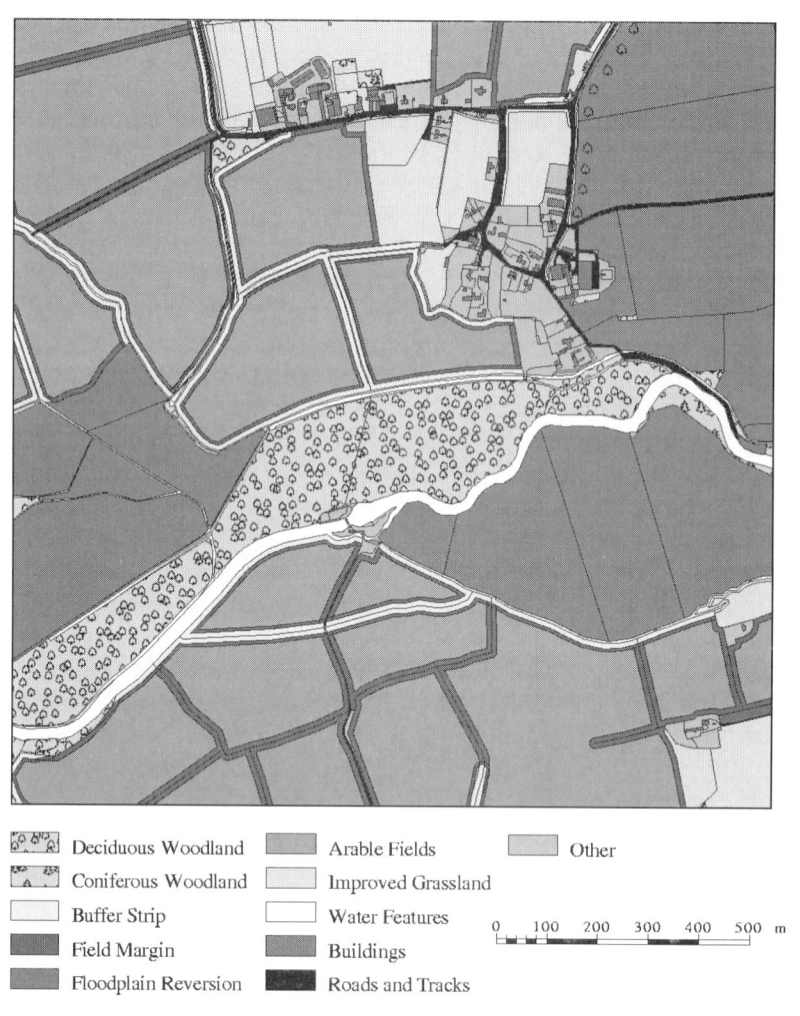

7.2 Land cover around Kelmscott in the Biodiversity Conservation scenario

Legend: Deciduous Woodland, Coniferous Woodland, Buffer Strip, Field Margin, Floodplain Reversion, Arable Fields, Improved Grassland, Water Features, Buildings, Roads and Tracks, Other

Map based on Ordnance Survey Land Line and Panorama digital data. © Crown copyright.

It was thought important to supplement the A0 maps with three-dimensional visualizations because the latter provided a much better sense of the implications of different scenarios for views across the landscape. Several three-dimensional visualization tools were investigated, but an approach based on VRML was ultimately selected on the following grounds:

- practicality of generating visualizations from within a GIS;
- scope for allowing interaction and movement within a landscape;
- potential for dissemination via the Internet;
- limited software costs.

VRML is an open standard for three-dimensional multimedia and shared virtual worlds on the Internet (see http://www.web3d.org). A VRML 'world' consists of one or more files (conventionally with a .wrl suffix) that together describe the geometry and attributes of objects in a three-dimensional scene. VRML files consist of standard ASCII characters, but when viewed in a web browser configured with a VRML interpreter (e.g. Cosmo Player, see http://ca.com/cosmo/) the text commands are processed to draw the defined objects. Facilities to generate VRML now exist in many GIS packages (e.g. ESRI ArcGIS or ERDAS VirtualGIS), but in our research we used the Pavan tools (Smith 1997; Smith 1998) which operate within the MapInfo GIS software.

The process of producing the VRML models is discussed in detail by (Lovett *et al.* 2002), but it is worth noting that during the course of the work several compromises became necessary. Most of the buildings were deliberately kept simple with a single colour, a standard height and a flat roof. It was technically quite possible to vary building heights or roof styles, but there was no data that would allow this to be done in a comprehensive manner. Another option would have been to place bitmap images or vector drawings on the sides of buildings, but a pilot investigation soon revealed that this substantially increased the processing and rendering requirements.

Simple tree and bush models were added to each landscape by first specifying the characteristics of the required features (e.g. type, height and colour), and subsequently positioning them using mouse clicks. There were no facilities in the authoring tools to cover polygons with set densities of trees, and so defining areas of woodland proved a rather laborious operation. Other difficulties arose in the visualization of field boundaries and this was eventually achieved using variations in line thickness and the density of bush symbols. The modelled representations were consequently stylized and more visually transparent than the real field boundaries.

Assessment of the visualized landscapes

Figure 7.3 shows the appearance of a VRML model when viewed using Explorer and Cosmo Player. The controls towards the bottom of the Explorer window provide facilities for using the mouse to move around the VRML world in various ways (e.g. zoom in/out, pan, slide, rotate and tilt), so it is possible to explore the virtual environment in a flexible and interactive manner. Specifically defined viewpoints can be selected from a box on the control panel. It is also possible to link several viewpoints in a sequence and generate a simple animated tour through a VRML scene.

Figure 7.3 shows a view across the current landscape looking north across the area mapped in Figure 7.2. The upper River Thames

Applications

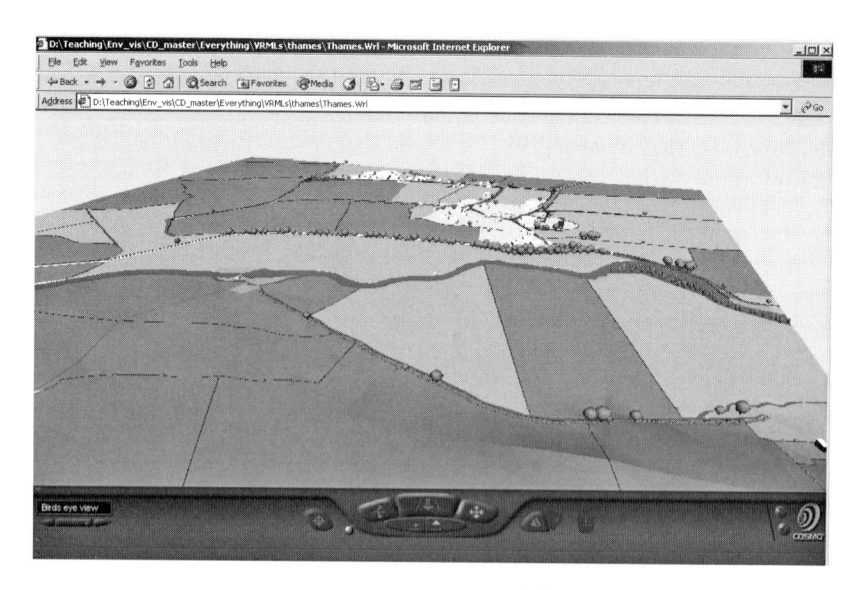

7.3 View of VRML model around the River Thames

7.4 View of VRML model for land cover in the Biodiversity Conservation scenario

dominates the centre of the image and Kelmscott village can be seen beyond it. Lines depicting the boundaries around the arable fields are evident on the clay vale, although individual bushes are difficult to detect against the dark background.

Figure 7.4 presents a view from the same point across the Scenario 3a (Biodiversity Conservation) landscape (i.e. it matches the land cover shown in Figure 7.2), and here the introduction of features such as buffer strips, field margins and enhanced hedgerows results in a substantial visual contrast. Other alterations include new areas of woodland beside the River Thames and the reversion of floodplain zones to rough grazing or marshland. Several narrow or small features (such as field margins and buffer zones) were deliberately given less realistic, but bright colours (shown as lighter shadings on Figure 7.4) so that they would be clearly visible. This was important for visualizing potential changes to the landscapes and also allowed elements of the VRML models (where there was no legend) to be related to the A0 colour maps.

One distinctive characteristic of VRML is the ability to interactively navigate through a landscape using the browser controls and this feature proved valuable during demonstrations of the research. It is fair to note, however, that even with the VRML files stored on the hard disc of a computer (rather than being accessed over the Internet) the speed of response to the controls was often rather slow. For demonstration purposes it was ultimately found easiest to define several viewpoints and an animated tour in each model. The latter was especially useful as it could be left to run for several minutes while members of the research team provided a commentary on what was being displayed.

A meeting attended by representatives from a variety of stakeholder organizations (e.g. English Nature, the National Farmers' Union, and the National Trust) was held in May 1999. The large colour maps and VRML models (projected from a laptop) were presented to the participants and they were asked for their comments on these, and other, aspects of the research. Reaction to the maps was very positive, although it was noted that they needed careful study to compare the present situation in a particular locality with the different scenario outcomes. Opinions on the VRML models were more mixed. It was accepted that they complemented the maps and had the potential to provide effective overviews of areas, but it was also thought that additional details and textures were necessary to make specific locations immediately recognizable by local residents. Other comments included the need to improve response speeds so that navigation around the landscapes could occur in a smoother fashion.

Final visits to the farmers were conducted in June 1999, 18 months after the initial questionnaire surveys. Only 17 farmers were interviewed in this phase of the project, but care was taken to include all those that would be most affected by the possible floodplain reversion or who initially had been most hostile to participating in whole landscape management. In each interview, the farmer was presented with the scenario maps, VRML landscape representations and flash cards summarizing key ecological benefits of the scenarios. They were also shown a summary of the likely land use changes on their farm under the scenarios. Their reactions to the prescriptions and their willingness to cooperate in landscape-based planning and management was then explored through a discursive, open-ended interview. Each interview took one to two hours.

The overall response surprised us: none of the farmers interviewed were strongly opposed in principle to redirecting their farm management to a comprehensive whole landscape. In addition, all the farmers said that the visual images helped them significantly to form their views on the scenarios, even those who were not willing to cooperate beforehand. It is important to recognize that the responses of farmers may have been significantly influenced by the deterioration in farm incomes, that options such as the restoration of floodplain grassland could not be achieved without long-term financial support, and that careful negotiation would be required in the implementation of such schemes (O'Riordan *et al.* 2000; Dolman *et al.* 2001). VRML also has some limitations as a means of creating landscape models and more recent work has examined the merits of other approaches (Appleton *et al.* 2002). Nevertheless, the positive attitude to the whole landscape proposals at the conclusion of the study can be hailed as a successful demonstration of the benefits of using landscape visualization techniques within a participatory decision making framework.

Acknowledgements

The Oxfordshire research was primarily funded by the ESRC Global Environmental Change Programme (Grant number L320253243). Additional support was provided by the Arkleton Trust, the Ernest Cook Trust and the Esmee Fairbairn Charitable Trust. Thanks are also due to the National Trust for their cooperation and to the Ordnance Survey for the provision of digital map data. These digital map data are reproduced by kind permission of OS Crown Copyright.

HELPING RURAL COMMUNITIES ENVISION THEIR FUTURE

Christian Stock and Ian D. Bishop

Envisioning systems

In many areas of Australia there are pressures for land use change. In the mountains of south-eastern Australia, livestock production has been the traditional dominant enterprise, but there is pressure for change as tourism and lifestyle development push land prices higher or forestry becomes more economic. In other areas of the country the long established heavy use of irrigation for rice or other crop production is looking increasingly unviable. As a result, both the structure of rural communities and the visual landscape are expected to change markedly in the next 20 years. These changes should not be driven by economic necessity without consideration of both local and wider community values, that is people's expectations for the visual character of the region and concern for its environmental health.

This section describes an Envisioning System (EvS) designed to help rural communities contemplate landscape level changes. Simulations and models project current conditions into the future according to the constraints of scenario-based planning and available land use choices. Possible future conditions are represented through visual (two-dimensional, three-dimensional and iconic) indicators. The goal of an EvS is to help community members negotiate desired future conditions and implement policies which shape land use changes that produce these desired conditions.

We have developed a portable EvS for landscape simulation and exploration. Our EvS is based on four components: geographical information systems (GIS), virtual reality (VR), physical process (impact) models, and mobile computing devices. In a community workshop, our environment will allow community and stakeholders of a study area to propose alternative land cover configurations in the GIS, move in real time through realistic three-dimensional renderings of the consequent landscape, and review a variety of environmental, economic and social outcomes. The community and stakeholders can then express their opinion on the outcome of the proposed landscape changes.

A virtual three-dimensional model of the existing features of a study area has to be produced as a first step (Chen *et al.* 2002). This must include a terrain model and three-dimensional objects representing the existing features, such as buildings, trees, fences and roads. The terrain is textured with orthophotographs. New land cover objects can be placed onto this base. It is also possible to remove existing three-dimensional objects, i.e. in general it will be desirable to remove exist-

Applications

ing trees if, for example, a pine plantation would be 'planted' onto the same management unit.

Simulation environment

The developed simulation and exploration environment is illustrated in Figure 7.5 (see also Stock and Bishop 2002). The landscape simulation environment is rendered in real time onto three screens using a single computer with three output channels, and so to three projectors to give a 135° field of view. This computer is also used as a server to control the network data flow and to send appropriate messages to the PA visualization program. This (master) computer is linked with another computer running ArcGIS, which will supply land cover information to the rendering system. Several PDA devices can also be connected to the master computer using a Bluetooth wireless network.

Our program called PA (Object Animation with Performer) is based on OpenGL Performer. This renders the three-dimensional representation of the landscape. PA can work with 'dynamic' objects, which can be animated (e.g. move along a path) or be switched between different states (such as 'on' and 'off') using triggers and targets. The extended functionality of PA is achieved by the use of 'scene files'. A scene file defines the interactive properties of the objects that will be loaded into PA.

Data exchange between PA, the GIS and the PDAs is all handled by a program called PA-server. The server is able to receive messages from multiple devices, e.g. the ArcGIS client and multiple PDA devices. The

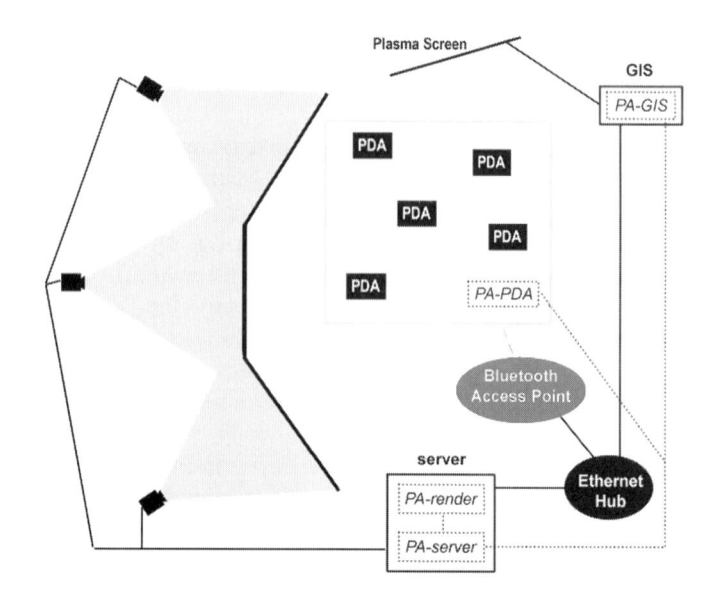

7.5 The simulation environment. PA is our three-dimensional visualization program, PA-server handles the communications between components and PDA stands for Personal Digital Assistant which communicates via the wireless hub. Three-dimensional graphics are displayed on the wide screen via rear-projection and the GIS interface is displayed on a large plasma panel

server understands several different message types for different tasks. Using ArcGIS it is possible to change the current land cover or to change the view position. The PDA devices can be used to update the view position, or to express an opinion on a specific land change outcome. The server is listening for all incoming messages in a multi-threaded environment, i.e. all messages sent from other devices will be received, even if they were sent at the same time, and be handled accordingly. The server will then pass the messages on to PA which will render the changes in the next frame.

User interaction and feedback

While the VR view is used to visualize the virtual landscape, the GIS and PDA interfaces are used for user input. Although the GIS is also used to show the study area in a two-dimensional view, its main use is to allow the users to manipulate the virtual landscape. We also allow the users to navigate around the three-dimensional landscape in real time using the GIS and PDA interfaces. Finally, we use the PDA interface as a platform for feedback on alternative land configurations.

Land cover manipulation

While it may be advantageous to be able to change the landscape arbitrarily in space, this approach would impact computing times to a point where it would no longer be real time visualization. Instead, the study area must be divided into 'management units'. Each management unit is a polygon and can have, at any one time, one land cover type. Existing maps can be used for generation of management units, for example, property boundaries, streams or land capability zones. Another option is to ask stakeholders to define their own management units by marking them on a map or aerial photograph.

Our plug-in for ArcGIS, developed using ArcObjects, allows a user to select one or more management units and change the land cover from a predefined list (see Figure 7.6). Whenever land cover is changed, the relevant management unit ID and the new land cover type is sent to PA-server and the VR view will immediately show the change.

We also allow the existence of multiple management unit layers. We refer to each layer as a 'scenario'. This way we can pre-build certain scenarios that may be of interest to the community. This feature can also be used to save certain scenarios as they are created during a workshop. The ArcGIS plug-in can select a scenario and send the corresponding land covers to PA-server. This way we can quickly compare the visual aspect between different scenarios.

Applications

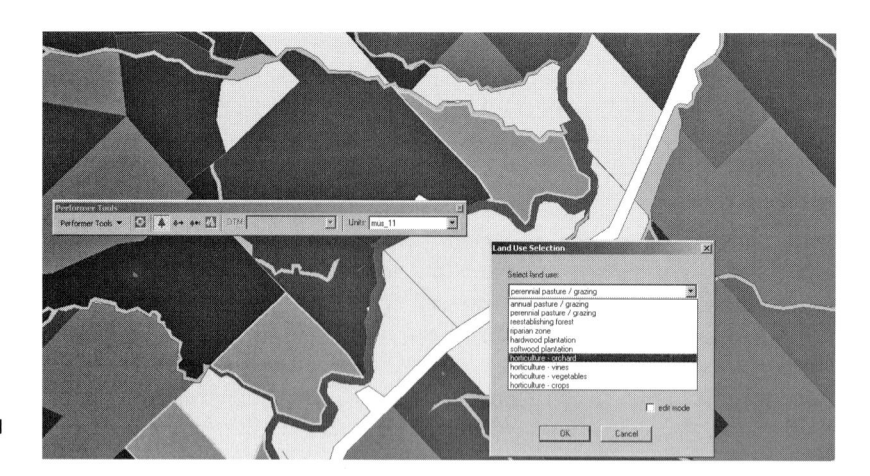

7.6 The GIS interface. Shown are different management units and the land cover selection box

To visualize different types of land cover, we use three-dimensional models that represent features typical for that type of land cover. If we want a management unit to represent forest, we would fill that unit with trees. To render the landscape in PA we use a base model (e.g. terrain with existing buildings), and sets of three-dimensional models for each land cover option and management unit, which are used to represent the predefined land cover options. Whenever a management unit is changed in the GIS, the appropriate predefined three-dimensional model is loaded into the scene graph.

In a workshop we will typically have one user controlling the GIS interface. The two-dimensional view will include the management units and other features like roads and buildings for orientation. During the workshop the participants will suggest land use change and the GIS user will change the management units accordingly. Having just one person changing the landscape avoids simultaneous changes that may confuse the meeting.

Navigation

The GIS interface can also be used for navigation in the VR view. The GIS user can click onto any point on the two-dimensional map and the viewpoint in the three-dimensional view will be updated. Another option is to select a location from a list of predefined viewpoints. All workshop participants are also able to select a viewpoint from the predefined list using the PDA devices. This allows a greater number of community members to navigate around the landscape – but some control is necessary.

The PDA interface also allows the workshop participants to 'walk' around the virtual landscape. The screen of the PDA is divided into quarters, i.e. top, bottom, left and right. Holding the stylus on the screen will result in continuous movement in the three-dimensional view, e.g. holding the stylus onto the top part of the screen will result in continuous forward movement. The other quarters will result in backward movement and left and right turns. We have elected to use only this simple navigation model at this stage because in order to achieve smooth, continuous movement in the three-dimensional view a lot of information has to be sent over the network, i.e. the current viewpoints have to be constantly updated. To limit network traffic the preferred option is predefined viewpoint selection or viewpoint setting on the GIS interface. The continuous movement functionality on the PDA devices should be used as a supplementary possibility to vary the view a little at a given location, but not to move over large distances.

Voting

Feedback from the workshop participants is an important feature of our approach. Those present can, of course, express opinions verbally at any time. However, there are also occasions when a more formal expression of preference is appropriate. This can be done using the PDA devices. At a given stage a question can be initiated via the GIS interface. The question is typed in and sent to the PA-server and distributed to each PDA device. A vote dialogue will appear showing the question and giving possible answers. Each member of the workshop will then be able to enter his or her preference. Once the voting is finished the results are sent back to the PA-server and then displayed on the VR view. The results will also be written into a log file for later evaluation.

Our system currently supports two types of votes. The first is a simple 'yes/no' question. For example, is the currently shown land cover scenario a viable option for the study area? The choice is 'yes', 'no', and 'undecided'. The second type of vote seeks a rating from one to five. For example, how sustainable is the current scenario?

The number of PDA devices is likely to be smaller than the number of workshop participants: people will have to share the PDAs. Each participant will vote and then will pass the PDA on to another person. Once everybody has voted, a 'done' button will be selected on each PDA device. The voting procedure is based on an honour system, i.e. it relies on workshop participants only voting once. Once every PDA has recorded that the voting is finished, the results are sent to the server. The server will evaluate the results and send the result to the main VR view. This will either be the proportion of yes/no votes or the average rating value. In this way the workshop participants will receive instant feedback and can discuss the results further if they so desire.

Applications

Impact models

The result of any land use change is not only visual. The quality of life of the local people may be changed in other ways through effects on their environment. Their concerns may include water quality and quantity, salinity, weeds, soil acidity, soil erosion and population numbers. Our system provides an immediate estimation of the non-visual impacts of landscape changes. For each predetermined indicator we can run a model on the GIS that estimates the comparative effect of any changes.

Since our system runs in real time we cannot use very complex models. A computing time of ten seconds may be acceptable but anything above that will stretch the patience of the audience. Thus, we cannot produce very accurate models, but we can give a correct general indication. The outputs are not absolute. Our models will indicate if a scenario increases or decreases each indicator relative to present conditions, or to some other scenario. For example, a scenario using extensive grazing may produce better water quality than a scenario using intensive irrigated horticulture. While for most impact models we will have slightly different ways of computing the values, the common approach will to associate a certain value with each land use type for each impact of interest. Those values will be determined from talking to experts in the respective fields. Finally, those values will be integrated in one way or the other to give a global estimate for the whole study area.

At any time the GIS user can select a tool on the GIS interface to compare two management unit layers (scenarios). The user can then select the impact models to be compared. The GIS will compute the relevant values and send the results to PA-server for display on the VR view (Figure 7.7). For each impact model we have a small two-dimensional icon that is displayed on the VR screen. The icon will be coloured according to the difference between the two compared scenarios. If there is a change to a larger value the icon will be bright green or dark green (bright green being the stronger change). If there is a change to a lower value the icon will be red or orange (with red being the stronger change).

The impact modelling combined with the visual assessment, through the VR view, should allow users to quickly assess any scenario. Simplified impact models could occasionally give a false indication. We accept this as a trade-off for real-time performance. More complex process models may produce greater accuracy but these are never accessible in a workshop environment. In seeking to provide people with an interactive tool for exploring and envisioning the future some analysis is better than none. A more thorough analysis should be undertaken on preferred scenarios.

Applications in the agricultural landscape

7.7 Screenshot of a virtual landscape with a proposed farm forest in the VR view. On the bottom of the screen are a compass which will help the users to navigate, a recent vote result (6 yes, 5 no) and six impact model icons (water quantity, water quality, erosion, salinity, acidity and habitat) (image reproduced in the colour plate section)

Outlook

We have developed a landscape simulation model that allows the users to explore alternatives to the existing landscape in an interactive, immersive virtual environment. This is achieved by using GIS technology as an input option for land cover changes that will be rendered in real time onto an immersive three-dimensional view. Using PDA technology, users can explore the alternative landscape environment, and assess and express their opinion on the outcome. We have also implemented simplified models to determine probable consequences of the landscape changes (e.g. jobs, amenity, water quality and quantity). Those non-visual outcomes of changes in land use will be visualized with the help of indicative icons. People can then determine if the probable changes accord with their values and adjust the extent or nature of the change accordingly.

Acknowledgements

Land and Water Australia (UME-65) funded this work. Andrew Kudikczak was the programmer for PA-server software and much of the rendering development.

LENNÉ3D — WALK-THROUGH VISUALIZATION OF PLANNED LANDSCAPES

Philip Paar and Jörg Rekittke

The efficiency of landscape planning projects, in both public participation and realization, is often negatively affected not just by various economical and social conflicts but also by cognitive limitations. In practice, attempts to implement landscape concepts often encounter resistance, if not obstruction, due to the difficulty of conveying the long-term, future advantages of the planning concept to decision makers and stakeholders. This problem has its roots partly in the inherent systematic complexity of the landscape planning process and also in the landscape itself. In particular, it is due to inadequate communication of the planning to stakeholders and the general public. While experts involved in planning may rely on their professional experience to help them visualize the landscape they are proposing, laypersons are often swamped with abstract and graphically insufficient maps and other landscape representations. They do not have the landscape experience for ready interpretation. Moreover, planning contents are seldom presented in a coherent and stimulating manner.

For some time now it has been possible to generate photorealistic landscape images using a personal computer (PC). However, such software applications involve a high degree of preparatory work and render time, and neither match the special needs of landscape planning nor exploit the technological possibilities of computer graphics to the full. A feasibility study in 2000 (Paar 2003) highlighted the need for an innovative new three-dimensional visualization tool. The planning practice is not content with the representational quality of plants and biotypes offered by available software solutions. Most landscapes are covered by vegetation; therefore 'convincing representation of plants and habitats' is a prerequisite for realistic visual simulations of the scenery, and a highly demanded feature for next generation landscape visualization systems. Since May 2002, the subsequent project *Lenné3D* (www.lenne3d.de) is addressing this challenge, making use of cutting-edge technology from developments in computer graphics as well as computer games techniques to enhance civic participation in landscape planning and design. Among those contributing to this collaborative, interdisciplinary research project are several computer graphics scientists, landscape planners and landscape architects, as well as a biologist and a crop scientist.

Aims of *Lenné3D*

Under the name *Lenné3D*, this collaborative project is involved in developing and testing an interactive system capable of visualizing spatial data, in particular the data of landscape plans, the scenery and the vegetation, both almost photorealistically from eye-level view and cartographically from bird's eye view. An eye-level perspective is termed 'first person view' in the world of computer games. For our purposes we will refer to *first person view landscape visualization*. The basic premise of the project is that computer-aided graphics can make the process of landscape planning and design more stimulating and transparent, as well as helping to overcome cognitive limitations.

The outcome of the project will be a prototype PC software application that is capable of generating landscapes models using data from Geographic Information System (GIS) and special algorithms, and visualizing these landscapes in real time. The emphasis of the three-year project is the realistic representation of vegetation as well as real-time rendering and the possibility for interactive exploration and the editing of spatial data. *Lenné3D* enables members of the general public, as well as professionals, to examine and explore landscape data, and additionally supports the planning and design process with interactive editing functions in three dimensions, thereby generating interest and ideas among the general public. The software enables decisions to be tested virtually by exploring visualizations of the resultant scenery.

The dedicatee of our project, Peter Joseph Lenné (1789–1866), stands for landscape architects and planers who are professional users of the *Lenné3D* system and who can creatively control the planning and design process. Lenné's name and repute are particularly auspicious for our purposes – his high-quality design graphics using three-dimensional effects were designed for easy visualization, conveying to the observer a particular spatial impression. We pay tribute to his visualization skills, which seem conventional only by today's standards, and are furthermore of the opinion that computer-simulated visualization should not replace but supplement traditional forms of representation.

A walk through *Lenné3D*

Below, we take landscape planning at community level as a notional example for the application of the *Lenné3D* system and describe the way it fits into the process in eleven stages (see also Figure 7.8).

1. *Involvement*. The community council makes a decision on the preparation of the local landscape plan. The head of local environmental or planning department is not the executive landscape

Applications

7.8 The landscape planning process facilitated by *Lenné3D*: (a) involvement; (b) the initial state; (c) distributing the plants; (d) the first person view; (e) editing in 3D; (f) walking the landscape

planner, however, he is someone with an understanding of the administrative, legal and political issues behind the project. He has goals, a vision or ideas, yet he needs the expertise of landscape planners to implement them. Usually, a private planning office is hired. The hired landscape planner uses the *Lenné3D* system for communicational purposes, visual analyses and design processes.

The computer expert – an assistant of the executive landscape planner – has complete command of the complex *Lenné3D* system. He conjures up the virtual landscapes for all those involved in planning to explore cooperatively, and is therefore one of the most important members of the *Lenné* team. His skills have not until now been a standard fixture of landscape planning offices.

Others are also involved in the process: lobby groups, the curious, the young and the old. Without their involvement, landscape planning can seldom be realized effectively.

2 *Beginning*. Regardless of whether the contract is the result of a competition, advertisement, or a direct negotiation, the start of every planning and design process is marked by detailed exchange of information between the landscape planner and the client. It is of particular importance at this stage that comprehensive landscape data is passed on to the landscape planner. The planner often does additional mappings. The more precise the data, the more potentially accurate the three-dimensional landscape visualizations will be.

3 *Initial state – GIS*. A standard tool used by landscape planners is the Geographic Information System (GIS), which allows the profile of the landscape and the environment to be summarized and represented. Using GIS, the spatial data can be used to create digital maps. GIS provides landscape architects with powerful tools for storing, analysing and modelling the spatial data. The GIS themes necessary for three-dimensional visualization with the *Lenné3D-System* are a topographical map, a digital elevation model (DEM) and other thematic maps, such as a habitat map.

4 *Initial state – Lenné3D*. The GIS data are loaded into the *Lenné3D-MapEditor*. The *Lenné3D-MapEditor* forms an interface between the spatial data and the interactive three-dimensional visualizations; the landscape planner uses it to direct the individual visualization projects and representation settings. This software component unites the advantages of conventional modelling with the classical two-dimensional map, in a digital form. The digital model of the terrain is processed into an interactive, three-dimensional visualization on the basis of the GIS data. Additionally, three-dimensional objects such as buildings can be loaded and landmarks and labels can be added. The data are summarized in a *Lenné3D* project. The interactive three-

Applications

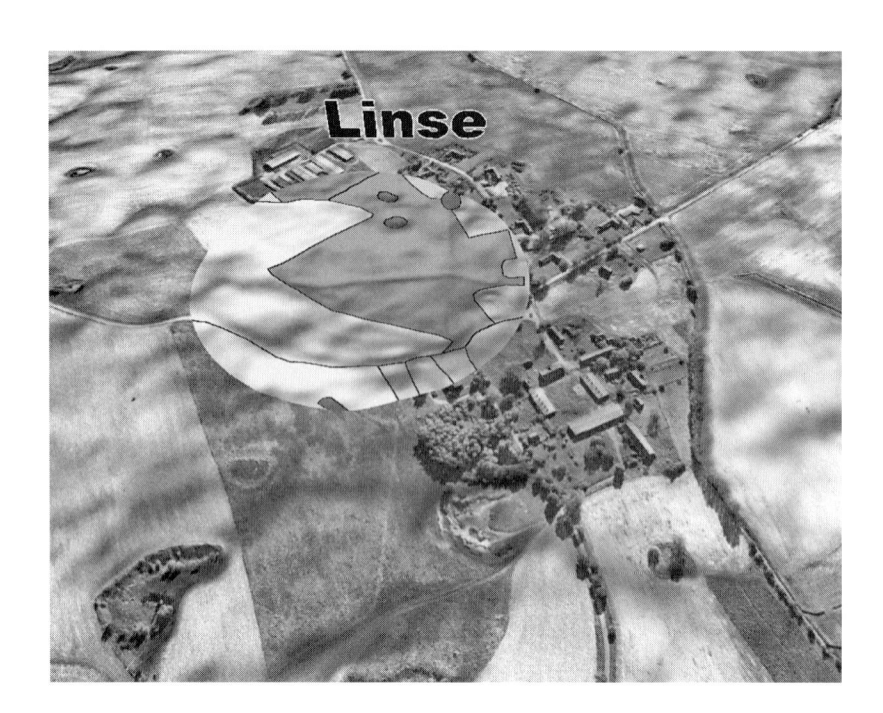

7.9 Screenshot from the *Lenné3D-MapEditor*; blending of an aerial photo and habitat map (ESRI Shapefile), both draped on a digital terrain model

dimensional maps provide an intelligible alternative to traditional maps, which demand a high level of abstraction. Yet they also use the clearness of traditional illustrations for editing or adding other geographical or landscape data. Special illustrative techniques such as *information lenses* allow a selective and clear communication of information (Figure 7.9).

5 *Virtual walk*. The clearness of the three-dimensional map representation in the *Lenné3D*-MapEditor allows those involved in planning to decide on a starting point or route for a walk through the landscape model, which they can explore as a photorealistic, virtual-reality model using the *Lenné3D-Player*.

6 *Plant distribution*. With the help of a further component called *oik* – *Lenné3D*'s vegetation modeller and plant distributor – GIS data are used to calculate rule-based distribution of plants for the most natural possible simulation of a real landscape. To this end, habitats are broken down into various types of vegetation. The plants are then distributed randomly within the vegetation layers and units. Mapped plants (e.g. a solitary tree) are positioned interactively or according to precise coordinates from a GIS point theme. Without this software component, the vegetation in the simulated landscape would be unnaturally regularly positioned. The landscape architect controls the *oik* component through the *3D-MapEditor*.

7 *Plant modelling*. The plant models used in the visualization afterwards are three-dimensional textured models, created using

the software *Xfrog* (Lintermann and Deussen 1998). They correspond to biologically correct criteria and, after conversion, allow a smooth level of detail and real-time representation. The user will have the option of selecting from a large number of ready-made plant models.

8 *First person view*. The *Lenné3D-Player* allows the observer to navigate through landscape section selected in the *3D-MapEditor*, interactively, in real time and with a breathtaking degree of realism, all from the perspective of someone walking through the landscape.

9 *Planning with GIS*. Following the inventory and analysis, discussions with the client and other stakeholders, and the preliminary walk through the digital landscape model, the landscape planner proceeds to develop a planning concept for the landscape. The objectives for nature protection and landscape management and of alternatives are saved as GIS data. Requirements and proposed measures for the client and stakeholders are derived from the planning concept.

10 *Editing in 3D*. The GIS data are processed for interactive use in the three-dimensional visualization by the *Lenné3D-MapEditor*. The distribution of plants is calculated for the types of habitats envisaged. The landscape architect uses the *3D-MapEditor* for analysis and the planning of modifications that will affect the scenery. The editing function with three-dimensional visual control allows different versions to be designed, all of which may be saved as GIS data.

11 *Walking the landscape*. Whilst walking through the interactive landscape with those involved in the landscaping process, specific suggestions may be taken on board; should there be aspects of the design that are unclear, it is possible to turn back for a second look, in order to gain a precise picture of the situation. Various different design options, planning alternatives or scenarios of landscape development can be shown successively or in parallel for the sake of comparison. Virtual measures can be selected or modified within the available decision space. If further modifications to the planning are necessary, the iterative process is repeated from stage 9.

Methods

Using real examples of landscape planning, the research and development team is engaged in testing application of the prototype system and developing the process of stakeholder participation supported by computer graphics. The first trial run started in early 2003 within the framework of the project 'Interactive Landscape Plan' in Königslutter (www.koenigslutter.de/landschaftsplan.htm), sponsored by the German Federal Office for Nature Conservation (BfN). The aim of the collaborative project is to test *Lenné3D* in practice under the conditions of a participative landscape planning process, and to evaluate its acceptance and results in an accompanying survey. In order to guarantee development of a system that will stand up to professional use, landscape planners are directly involved in the application design of the software. Furthermore, specialist seminars with potential professional users are conducted regularly.

Contrary to the linear software development originally envisaged, an iterative, incremental procedural model is the chosen method of approach. This model corresponds to contemporary standards in software engineering and is particularly suitable as a development tool for distributed projects with complex software systems.

The software architecture of the system designates two main components: the *3D-MapEditor* for the assembly, exploration and editing of landscape data, and the *3D-Player* for the interactive landscape visualization from first person view. This partitioning arises from the implementation of specialized and optimized computer-graphics systems, developed as separate projects, for each of the two components. Both of these main components employ the *oik* component, developed for the generation of plant distribution.

The *3D-MapEditor* is based on the authoring software for three-dimensional maps and three-dimensional city models *LandEx* (www.landex.de), which is being developed further for application in the field of landscape visualization. Analysis and editing functions, which serve to ascertain ecological parameters, have been integrated into the *3D-MapEditor* especially to fulfil the requirements of vegetation modelling. With the *3D-MapEditor*, *Lenné3D* projects are created and managed; the component can also integrate and transform geo-data.

The *3D-Player* is a program that interprets the scenic description and geo-data specified in a *Lenné3D* project. Three-dimensional vegetation models are assigned to the plant distribution calculated by the *oik* component and positioned within the landscape model. At the forefront of this technology is the development of the Level of Detail (LoD) process for the dynamic reduction of the overall complexity of the scenery. The geometric and colour complexity of the landscape model must be drastically reduced in order for the real-time simulation in the *3D-Player* to be possible, but this must remain imperceptible to the viewer. This requires that the relevant visual information is conveyed and packaged in suitable data structures for efficient data processing and quick access. The latest computer graphics techniques allow a large number of images to be generated and calculated per second. Triangles of the complex model are progressively replaced with fewer primitives (points and lines) as the camera moves away. Representing object surfaces as sets of points or lines without connectivity allows for easier simplification and generation of LoD representations. The point and line algorithm projects each point or line sample onto the screen and draws it as a pixel. At scene level, whole objects are progressively 'melted' into the terrain texture as the camera moves back.

The development and implementation of illumination procedures for plant and vegetation models involve some special features with regard to prior illumination methods. For example, in certain situations, the light is dappled as it shines through the plants or their leaves. The illumination model must fulfil this additional requirement. Special new graphic hardware and programming capabilities are utilized in the *3D-Player* for a flexible implementation of such illumination models, so that they can be adapted to various user requirements.

The *oik* component calculates plant positions by means of heuristic-algorithmic models, on the basis of habitat data (biotope types), reference mappings and topographical site data. In cooperation with the company Greenworks Organic Software, best known for its *Xfrog* software, a wide range of central European plant life, particularly foliage, has been modelled. Plants commonplace in central Europe are selected, and supplemented by species typical of the region in which the testing areas are situated. The three-dimensional plants are modelled for the *3D-Player* in full spring appearance. This decision was made for reasons of working capacity, as well as the fact May–June seems the most suitable phenological reference season for visual simulation, as the vegetation is most lush at this time. Woody plants are also modelled for a winter view.

First results

Initial stages of the project aimed to construct a basic framework for the development and integration of the software components of the *Lenné3D* system. The intention was to integrate the different, hitherto rudimentary, software components at an early stage of the project, in order to avoid a 'big-bang' integration in the final stages – which, in the field of software engineering, carries a greater risk than an iterative, incremental approach.

A prototype system is already being tested with GIS data sets. A preliminary version of the *3D-Player* can already be used to visualize smaller landscape scenes photo-realistically and in real time (Figure 7.10).

In collaboration with our partner project 'Interactive Landscape Design', a 1.1 km^2 slope was selected for high-detailed landscape visualization, in discussion with local stakeholders. The area, which is predominantly used for agriculture, is situated at the edge of the Elm Natural Park, Germany and is affected by erosion. The town wishes to enhance soil protection, promote landscape-related recreation and protect the natural species and habitats it houses.

Applications

7.10 Screenshots from the interactive *Lenné3D-Player*, geodata from the Uckermark district: (top left) existing; (top right) as planned with removal of farm building; (bottom left) and (bottom right) higher plant density, more species and more than 2.5 million three-dimensional plants instances (bottom images reproduced in the colour plate section)

Conclusion

The profession of landscape planning 'must be prepared to keep up to date with current developments in the field of digital technology and, if necessary, develop solutions tailored to its needs' (Rekittke 2002: 121). After two years of development, the project's goal of real-time visualization of complex landscape scenery is deemed realizable considering the performance that is already possible using current three-dimensional graphics cards, as well as the increasing performance made possible by developments in hardware and corresponding algorithmic innovations. Nevertheless, an effective solution has to be found for processing the considerable amounts of data involved in large and complex landscape scenes with millions of plants.

But how will landscape planners and stakeholders react to the new technological possibilities, especially in controversial situations? This new form of communication could stimulate interest in civic participation, while promoting a fun, educational approach to environmental protection – thereby generating more interest among young people and members of the population who have so far shown only scant interest. Three-dimensional landscape visualization is by no means a typical service of landscape designers. Until now it belonged to the domain of the film and computer-game industries. Yet the influence of new media technology is reshaping our visual habits, and along with these the demands on and potential for presentational illustrations of our environment. Among landscape planners and their clients there is increased awareness of the benefits of landscape simulation during the planning and design process, not purely as a medium for presentation, but as an integral and integrating component of a participative design process, in a departure from the classical, linear model.

Architects and town planners have always constructed three-dimensional models (either physical or virtual); now, landscape planners will also be able to use visual simulation to add a three-dimensional aspect to planning. Visualizations from first person view using GIS data are coming into conflict with the scale of conventional representations of community landscape plans. A greater degree of detail can call into question the accuracy of the basic data and the postulated planning scale of, for example, 1:10 000.

As more 'bottle-necks' are removed by the fast pace of developments in computer graphics, the more transparent typical issues of landscape design such as cost, availability, quality and incongruities of data and models will become apparent.

The research project *Lenné3D* has set itself high aims. In part they are so high that they cannot be completely fulfilled during the term of the project, yet it is already manifestly apparent that the research is leading to developments that will considerably improve future methods of communication and visualization during the planning and design process. Our strategy of holding bi-annual seminars with professional experts and users to hear their assessments and criticisms of latest developments in the research project has proven extremely valuable, as well as demonstrating that our results have already inspired confidence and enthusiasm among user groups.

It becomes apparent 'that landscape, by virtue of its inherent intellectual or virtual nature, is ideally suited to being experienced and conveyed through digital media' (Rekittke 2002: 110). However, an important revelation for all those involved in the project was the realization that the reality of natural, botanical and scenic beauty could not be even closely approximated through computer simulation. Rather than a source of disappointment, we see this as a continuing challenge for our current and future work.

In addition, the current rapid technological developments and our own development milestones have served to increase our respect for the skill of conventional methods of representation. Our chosen name

Lenné3D serves as a constant benchmark for checking and recalibrating the quality of our graphics.

Acknowledgements

The German Federal Environmental Foundation (DBU) sponsors the *Lenné3D* project. *Lenné3D* is a cooperative project of the Institute of Land Use Systems and Landscape Ecology (ZALF), the Zuse-Institut Berlin (ZIB), Department of Scientific Visualization, the University of Konstanz, Deptartment of Computer and Information Science, Chair Media Informatics, Prof. Dr Oliver Deussen, and the Hasso-Plattner-Institute for Software Systems Engineering (HPI), Chair Computer-graphical Systems, Prof. Dr Jürgen Döllner.

CHAPTER 8

APPLICATIONS IN ENERGY, INDUSTRY AND INFRASTRUCTURE

INTRODUCTION
Ian D. Bishop

Infrastructure proposals (roads, bridges, power-stations) have for a long time been drawn in three-dimensions using computer-aided design or similar modelling tools. While the objects themselves were drawn, their context was often neglected. The focus was on the engineering and not the environment. The purpose of early simulation work in this field was to impress the public with the scope of the investment and the wonderful advantages development would bring. It was only after the acceleration of environmental awareness triggered by Carson (1962), Ehrlich and Ehrlich (1970), the Stockholm Earth Summit of 1972 and the rising power of OPEC in the 1970s that attention began to turn towards the role of visualization in promoting the wider consequences of industrial development. The importance of the relationship between industrial development and the visual landscape also became increasingly crucial at this time. An early example was the preservation of the historic Hudson River landscape from power-station development as reported by Petrich (1979). By the end of the 1980s representation of proposed power lines, sewage treatment plants, railways and harbours (Figure 8.1) as new landscape elements was commonplace. Indeed, in Sheppard (1989) a large proportion of the examples are from the industrial/infrastructure context.

While in other contexts, visualization has developed additional capacities in terms of interactivity, links to GIS, on-line manipulation, use in community workshops etc., in the industrial context this trend

Applications

8.1 An example of the use of photomontage in an industrial setting as used at the Dibden Terminal public inquiry: (top) before; (below) after (images reproduced in the colour plate section)

has been less apparent. As the case studies of John Ellsworth (see 'Visualizing scenic resource impacts: proposed surface mining and solid waste sanitary landfill', p. 166) and colleagues make clear, the wide range of options which may exist in other contexts is not typical of the industrial development situation. The situation is more likely to be one of a specific project being proposed, simulated, reviewed and debated (e.g. the case of a strip mine). Sometimes a small number of alternative sites are available and visualization is used in the choice process (e.g. the case of siting a landfill). However, industrial development is seldom as multi-faceted as in urban design or agricultural contexts.

The hottest topic of infrastructure development in the first decade of the twenty-first century is, at least in the western world, the wind farm. The very rapid development of the renewable energy market in the past decade has generally outpaced the planning mechanisms required to support optimal (on multiple criteria) location of farms and individual turbines. As a consequence, a generally popular development has met considerable opposition because of insensitive approaches to siting. Frequently, coastal environments are not only the windiest but also the most visually sensitive. Battles are fought on aesthetic criteria as seldom before.

Now there arises the opportunity for a more interactive and discursive approach to industrial visualization. Turbine location is often less constrained by geology, slope, water-access or transport infrastructure than past developments in energy production. This gives the opportunity for more interactive community participation in the planning and design of wind farms. In their section of this chapter (see 'The provision of visualization tools for engaging public and professional audiences', p. 175), David Miller and colleagues report on the use of interactive techniques for review of turbine siting and their application in community meetings. Their report considers broader issues in the context of visualization for rural policy communication in addition to the specific case of wind turbines.

The high profile of wind energy development has generated a large body of reviews and simulations of turbines. John Benson (see 'The visualization of windfarms', p. 184) reviews this body of work and concludes that methodological problems still exist with visualization and its presentation.

This chapter also illustrates the wide range of software and display options still employed in the visualization business. As Ellsworth and colleagues argue, while siting of a specific infrastructure entity may be a data-rich situation (as object and context can be modelled in detail), a mine is a dynamic entity which changes as it develops in ways which may not have been predictable in the beginning. This leads to a data-poor situation and makes recourse to image manipulation software and professional interpretation of probabilities a suitable response. A large colour plot of simulated future conditions may be the most appropriate display. At the other end of this continuum, the wind turbine is a very precisely defined and easily modelled entity. Data-driven and interactive approaches with immersive presentation in a virtual environment become the preferred display modes.

The analysis by John Benson suggests that the mode of simulation presentation is more important than we have commonly considered it to be, with scale being very poorly communicated by standard reporting formats. David Miller's group examines different presentation options while also stressing the importance of movement in setting the visual context.

VISUALIZING SCENIC RESOURCE IMPACTS: PROPOSED SURFACE MINING AND SOLID WASTE SANITARY LANDFILL

John C. Ellsworth, Abraham N. Medina and Issa A. Hamud

Introduction

Aesthetic impacts are defined as changes in the fabric, character and quality of landscape resources as a result of development or modification and the effect of those changes on people (Landscape Institute 1995; see also Ellsworth 2001). These can emerge as a result of development pressure, changes in land management, or through changes in production processes. The significance of the impact is a function of the sensitivity of the affected area and visual receptors (i.e. user groups) and the magnitude of change they will experience as a result.

Computer visualization techniques can be effective for assessing visual resource impacts of proposed surface mines and for siting solid waste sanitary landfills. Although both surface mining and landfill change the natural terrain and thus have potentially significant visual impacts, there may be significant differences in terms of the potential to predict the detail of the change and its progression over time. This affects the tools which can be best applied for visualization of the changes. Case studies are used here to review options for visualization in the planning and design operations. A recent mining project illustrates the effective use of image processing. A solid waste sanitary landfill project (Medina 2003) added CAD-based computer visualization techniques to compare visual resource impacts at different sites. Following the case studies, legal, social and technical aspects of simulation development are reviewed.

Surface mining case study

The Barrick-Mercur Gold Mine is located in west central Utah, near the town of Tooele. This cyanide heap leach process mine is located near US Forest Service and BLM land. The mine operators contracted with Ellsworth and Associates, landscape architects, inc. (EALA) to produce a series of computer visual simulations of the proposed reclamation, erosion control and revegetation activities. Visual simulations represented proposed mine expansion activities and subsequent reclamation over a ten-year time period. EALA's landscape architects were asked to first design, then simulate a new landform for the heap leach area that would be visually similar to the natural landforms of the area.

Applications in energy, industry and infrastructure

8.2 Barrick-Mercur Mine panoramic view: (top) existing condition; (below) visual simulation after regrading and revegetation (images reproduced in the colour plate section)

8.3 Barrick-Mercur Mine north side: (left) existing condition; (right) visual simulation after proposed drainage and adjusted road alignment. Pockets of snow were simulated to show the shape and form of the proposed drainages

8.4 Barrick-Mercur Mine operations area: (top) existing condition; (below) visual simulation after post-mining landform and revegetation

EALA staff also developed the conceptual revegetation and landform design for the existing administrative and ore processing facilities.

Figure 8.2 illustrates the full extent of the new landform in panorama, and shows the potential for revegetation and erosion control on the slopes. The visual simulation shows the flat top of the leach area re-contoured and shaped to resemble the surrounding landscape.

Figure 8.3 shows a view of the north side of the landform. The proposed drainage and adjusted road alignment are also visible. Pockets of snow were simulated to show the shape and form of the proposed drainages.

A third visual simulation (Figure 8.4) was used to communicate the intent of the mine operator to leave the site with an acceptable landform, revegetated, and without erosion problems. The existing access road was conceptually designed, meeting grade at the upper and lower ends.

Solid waste sanitary landfill project

As part of a landfill siting study in Cache County, Utah, three sites were evaluated to assess their visual value and to determine their suitability as potential landfill sites. Each was evaluated using the USDI Bureau of Land Management's Visual Resource Management (VRM) System. This system was chosen due to its widespread use in the western region of the USA and its applicability to large-scale public projects on predominantly natural lands. As specified in BLM Handbook H-8400, Visual Resource Management (USDI 2002a), an inventory of existing visual resources was performed, followed by a contrast rating of the proposed project.

Landfill siting alternatives and descriptions were provided through Phase II of the Cache County Landfill Siting Study, prepared by HDR Engineering (2000) and the City of Logan's Division of Environmental Health. For each site, HDR Engineering provided conceptual landfill layouts (4/3/2002). These layouts were received as AutoCad DXF files and hard copy printouts. Additional information, including topography and planimetrics for the study area, was also provided.

An inventory of existing visual resources was performed using BLM Handbook H-8400-10, Visual Resource Inventory (USDI, 2002b). This process uses a combination of scenic quality, sensitivity level and viewer distance, to assign various levels of visual resource value. These range from Class I to Class IV, with Class I having the highest visual resource value and Class IV the lowest.

Associated with each class is a different management objective relating to its visual resource value. Class I areas, having the highest value, are managed to protect the existing character of the landscape. Proposed projects in these areas must show little change to the surrounding area and must not draw attention. Class II areas are managed to retain the existing visual character, and allow for minimal levels of change. In these areas, management activities may be visible, but must not draw attention. Class III areas are managed to partially retain the existing landscape character. Changes in this zone can draw attention but must not dominate the view of casual observers. Class IV zones are areas where existing visual resources are least affected by change. In these zones, management activities may draw attention and may be the major focus of casual observers.

1.4 Repton's approach to before and after landscape representation

1.9 The change from the representational styles developed in the 1970s to the far more elaborate and realistic landscapes of the 1990s: (top) landform using distorted squares; (top right) landform with orthophoto overlay; (above) the original tree representations; (right) highly detailed tree models

2.6 Three-dimensional landscape representation but with solid colours chosen to match those in corresponding GIS mapping

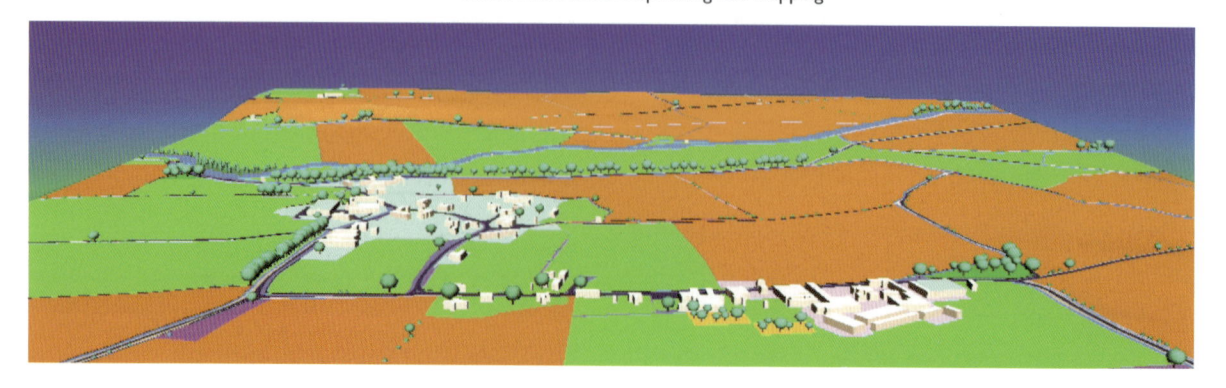

2.9 The distinction between geospecific and geotypical textures and objects: (left) geospecific ground texture and geotypical trees; (below) geotypical ground textures and housing

2.10 The three views of GIS data as defined by Verbree *et al.* (1999): (left) plan view; (middle) model view; (right) world view

4.4 A synthetic landscape modelled by Bernd Lintermann using xfrog and the open system described in Deussen et al. (1998)

4.7 Representation of (left) water with pixel shader; (bottom) water curtain on a plexiglass wall

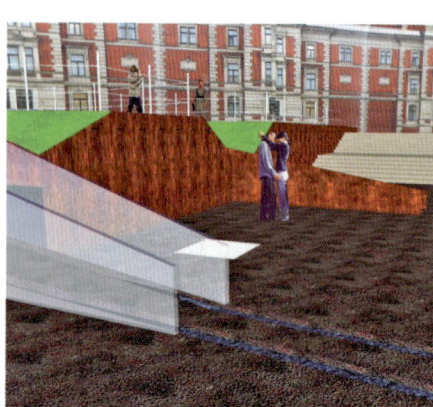

4.8 Interactive modifier for the fountain height

5.5 Sample frame from the Guanella Pass Proposed Road Improvements Project CD ROM, showing a drive-through video-animation in a user-selected driving sequence of the proposed highway improvements

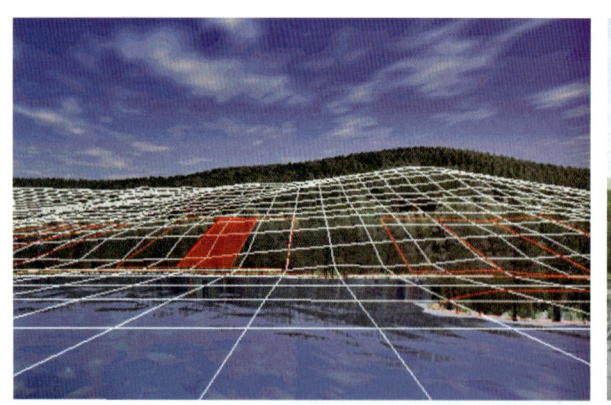

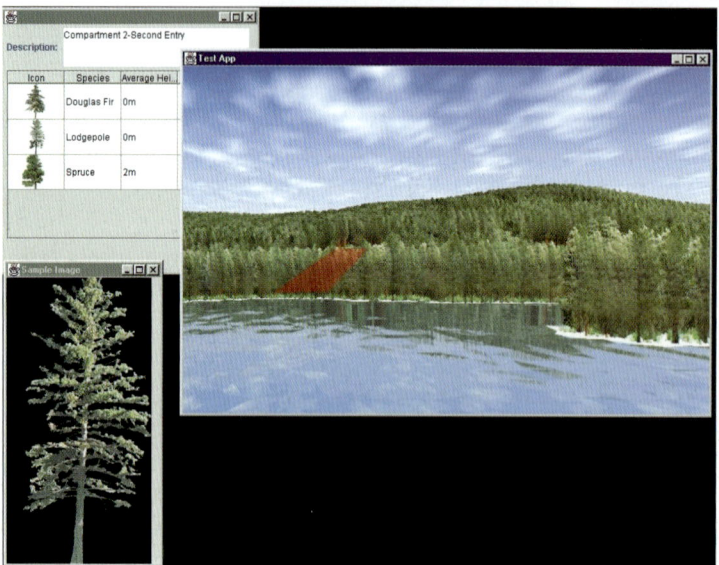

5.6 Making visualizations transparent – a simple Java application interface designed to display simultaneously both the modelled image and the underlying data: (top left) three-dimensional forest canopy wire-frame model; (top right) existing site photograph; (left) billboarded tree templates

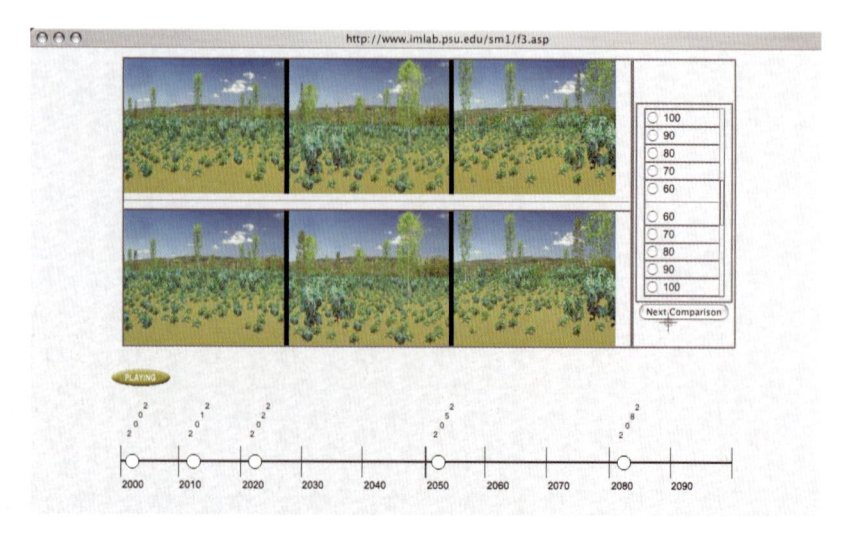

6.4 Typical survey page

6.8 Frames from the animation illustrating the dispersed retention harvest system: (top) immediately post-harvest; (middle) two years later; and (bottom) seven years later

6.10 Visualization of alternative forest management scenarios at Year 25, showing overlays of habitat modelling for one bird species

Forest Practices Code Scenario Year 25

Zoning Scenario Year 25

6.13 The Monsu-simulator can illustrate the effect of the different seasons on the landscape

7.7 Screenshot of a virtual landscape with a proposed farm forest in the VR view. On the bottom of the screen are a compass which will help the users to navigate, a recent vote result (6 yes, 5 no) and six impact model icons (water quantity, water quality, erosion, salinity, acidity and habitat)

7.10 Screenshots from the interactive *Lenné3D-Player*, geodata from the Uckermark district showing high plant density

8.1 An example of the use of photomontage in an industrial setting as used at the Dibden Terminal public inquiry: (top) before; (below) after

8.2 Barrick-Mercur Mine panoramic view: (top) existing condition; (below) visual simulation after regrading and revegetation

9.2 Overview: status quo

9.3 Agriculture scenario

9.4 Recreation scenario

9.5 Nature conservation scenario

9.6 Wind turbines scenario

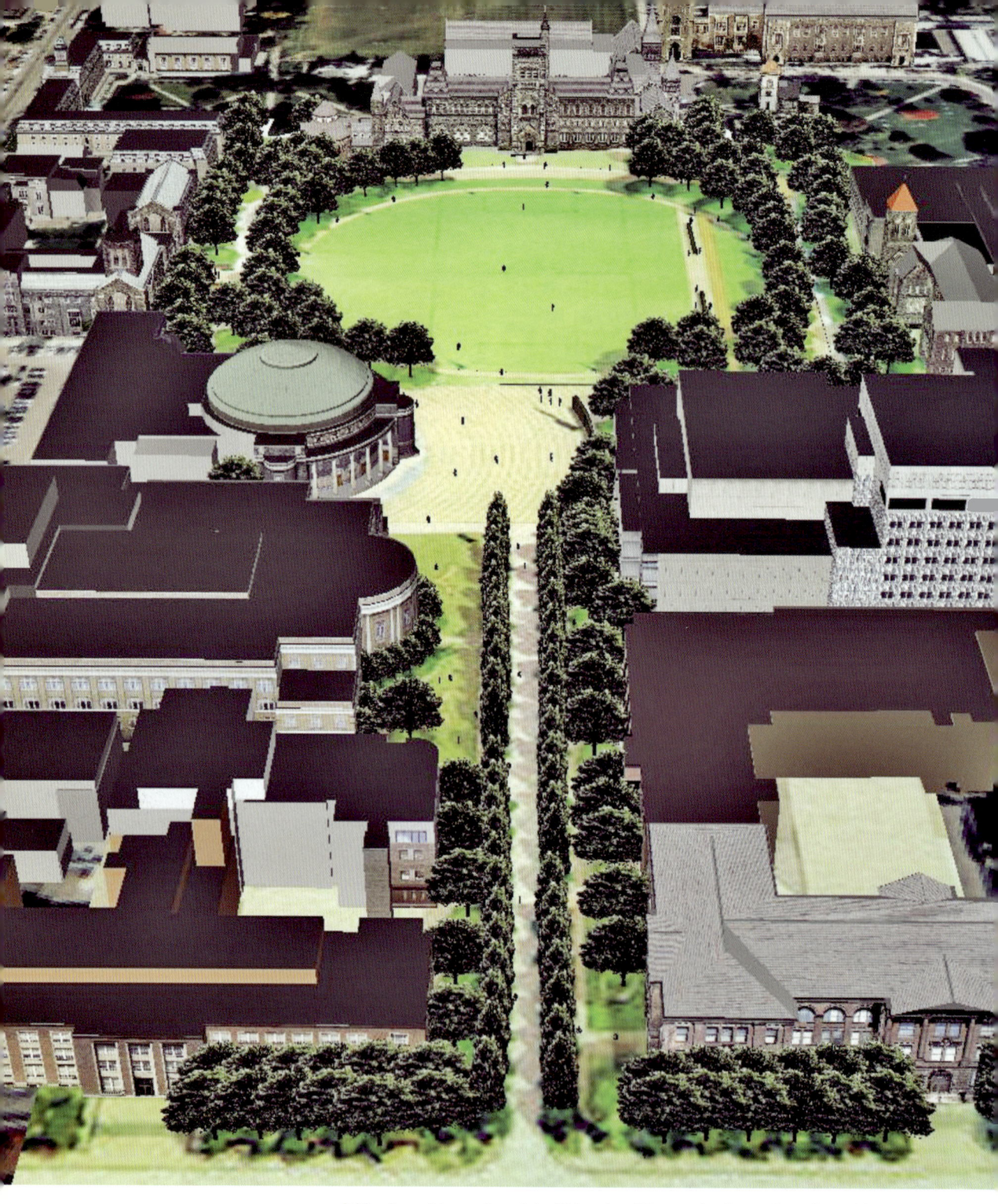

9.10 An aerial overview of the University of Toronto Landscape Open Space Master Plan Concept designed by Urban Strategies Inc. Models such as this one become useful if one can freely look about and move around at eye level, comparing visual concerns with systematic economic and planning parameters that affect feasibility and profit

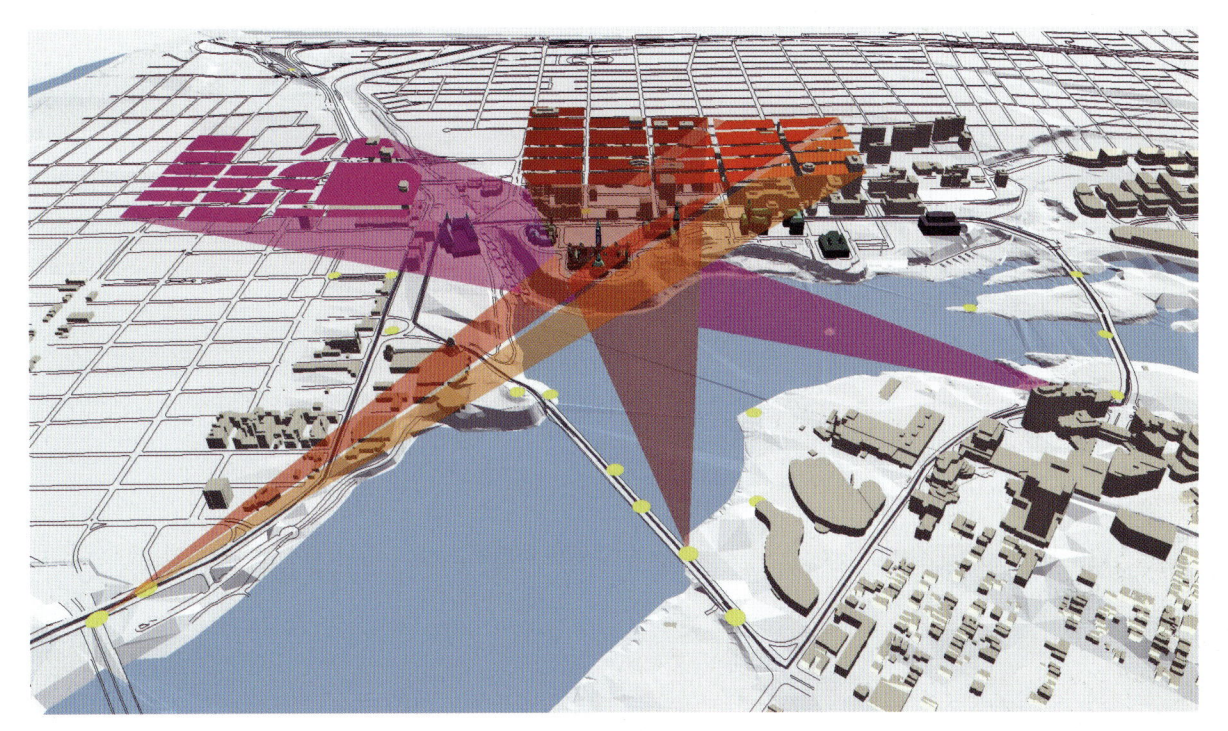

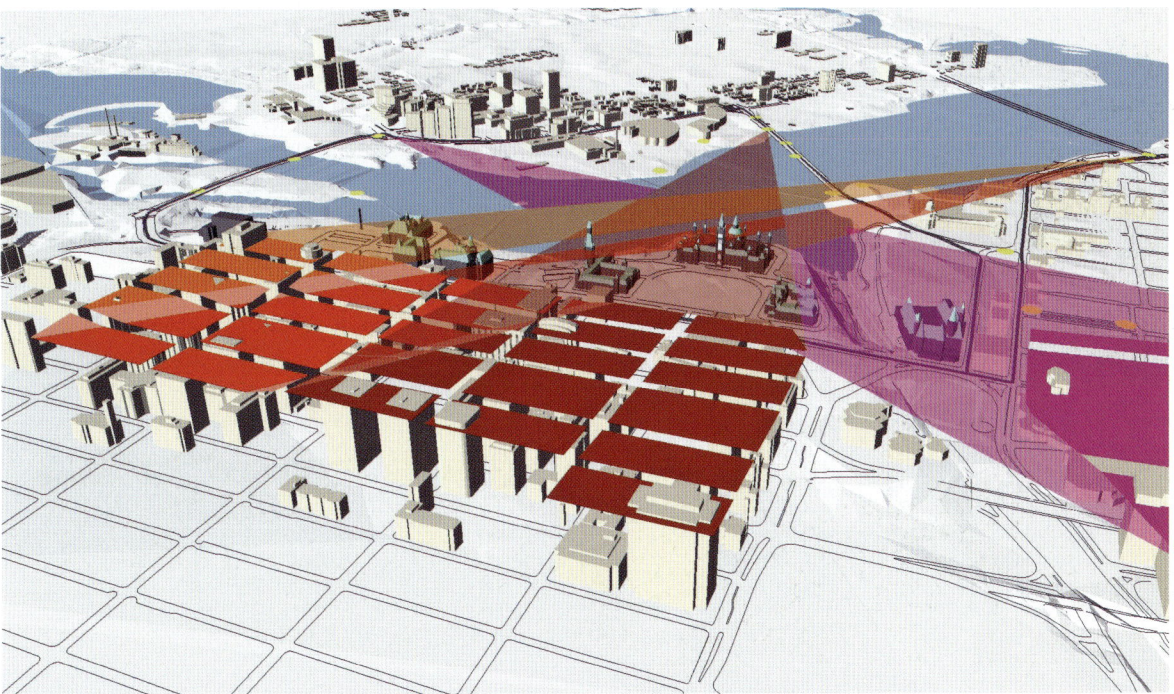

9.14 The controlling views selected as the systematic basis for controlling heights in Ottawa and the key viewpoints (yellow dots) along the continuous and dynamic views of the ceremonial and interpretive routes

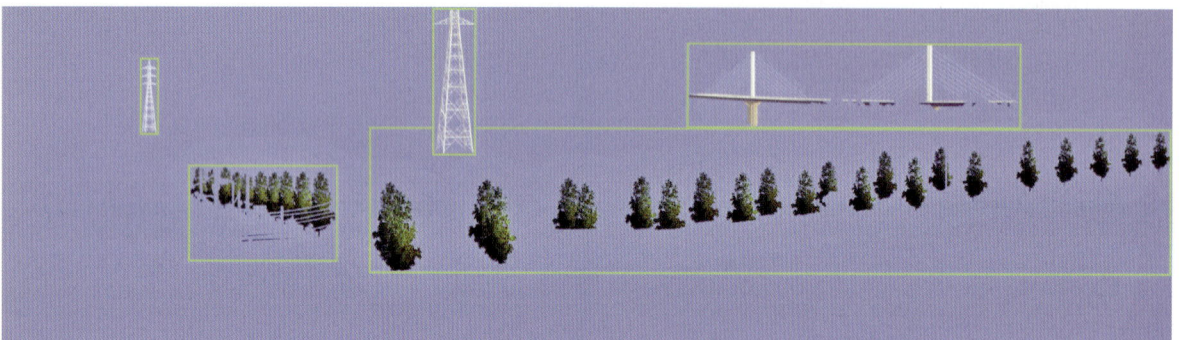

10.1 Panoramic scenes: (a) original panoramic scene composited from a panning video sequence taken in the evening; (b) panoramic scene of computer-generated images corresponding to (a); (c) composited panoramic image of (a) and (b); (d) panoramic image composited from a panning video sequence taken in the afternoon

10.5 Tinmith-Endeavour outdoor augmented reality mobile computing system

10.8 Augmented reality view of the new building

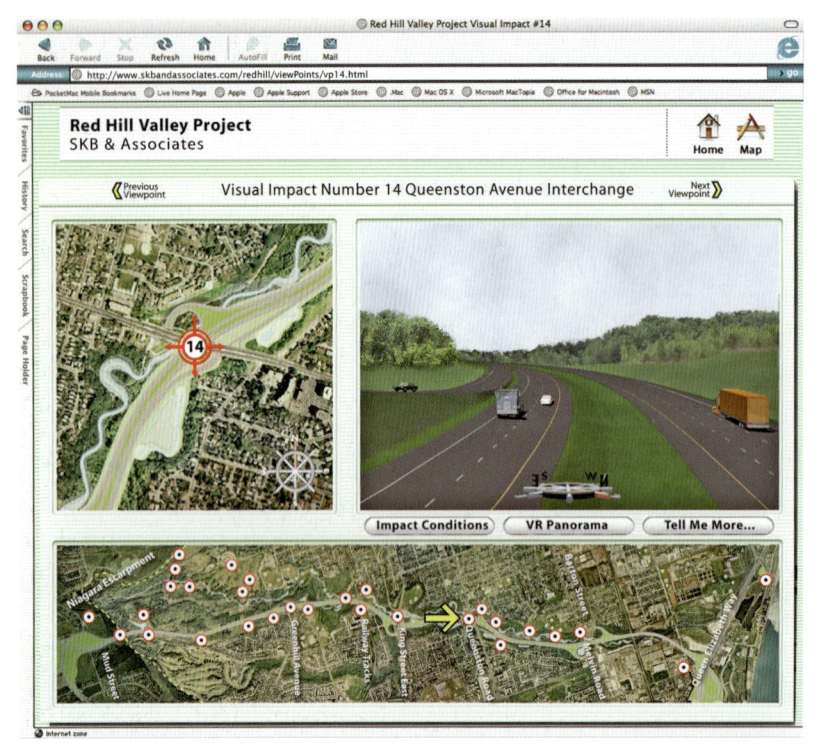

10.13 Red Hill Valley Main Navigation Page showing from top left and then clockwise: individual view locations with navigation arrows, 360° panorama and overall key map

10.19 Eye-level walk throughs: (left) existing condition; (right) redevelopment

Using known travel routes, user assumptions, public comment and professional judgment, Key Observation Points (KOP) for each landfill were identified and marked on USGS quads. These points represent locations in the landscape where project activities would be most revealing to the casual observer. In total, 13 KOPs were identified. BLM Manual 8431, Visual Resource Contrast Rating (USDI 2002c), was used to assess the visual impact of future landfills on three individual sites. Visual simulations at three time intervals were created for each landfill KOP, aiding in the contrast rating process. A total of 39 visual simulations were developed.

In order to determine the accurate visual extent and layout of the three proposed landfills, Autodesk 3D StudioViz R2 was used in the development of three-dimensional perspective views from each KOP. Schematic landfill layouts, showing size and form, and their exact locations within the valley were obtained from HDR Engineering in CAD format. This information was directly imported into StudioViz, where landfill and valley contour information was used to generate triangulated irregular network (TIN) models of the study areas (Figure 8.5). Accurate scenes, showing the height and horizontal extent of the landfills, were created as a result (see Watzek and Ellsworth 1994 for a discussion of accuracy in computer visual simulations).

Using the computer generated three-dimensional models for reference (Figure 8.5) photo-realistic visual simulations of each landfill were created using image editing software (Adobe Photoshop 6.5). Visual simulations were created for each landfill at completion, along with visual simulations at pre-selected time intervals (Figure 8.6). These were based on the predicted lifespan and proposed build-out plan of each landfill, as provided by HDR Engineering.

For each KOP, the existing landscape was considered in terms of its basic features (i.e. form, line, colour, texture) and basic elements (i.e. landform/water features, vegetative features and structural features), and the results recorded on a worksheet. Using the prepared visual simulations, each proposed project (at different time intervals) was also considered using the same basic elements and features. By comparing the existing and proposed features and elements, the degree of contrast between both scenes was measured. The project received a contrast rating based on BLM guidelines.

8.5 Site C as seen from KOP 13 showing the use of three-dimensional modelling as a guide for image-based simulation: (left) three-dimensional model of site from 3Dstudio Viz; (right) photograph of the site with three-dimensional model overlay

Applications

8.6 Landfill at (left) 10-year simulation; (right) 80-year simulation

The resulting contrast ratings were used to determine the project's compliance to visual management class objectives, assigned in the baseline inventory. By roughly correlating contrast rating (none, weak, moderate and strong) to the four visual management classes (I, II, III, IV, respectively), compliance was determined. Project sites that did not meet the area's visual management class objectives were identified, and additional mitigation measures were proposed. In some cases, mitigation was suggested for projects that did meet an area's management objectives, in order to further improve the project's visual appearance.

Simulation issues and process

Legal and procedural issues

The evaluation and assessment of impacts to visual resources in the western USA, especially those on federal government agencies' lands, often require an Environmental Analysis (EA) or an Environmental Impact Statement (EIS). These are detailed environmental analyses, with the EIS being the most extensive. Most US federal land management agencies have specific and detailed visual analysis systems for inventory, assessment and management of impacts to visual resources, including visualization as an analysis tool.

Two components of these visual analysis systems are very important to surface mining in particular. First, in order to determine visual impacts, or contrasts, of proposed surface mining activities the existing landscape's visual character, often referred to as the 'characteristic landscape', must be inventoried and described. With surface mining, the existing landscape sometimes includes previous mining activity which may, by virtue of historical precedent, actually be considered part and parcel of the existing 'characteristic landscape'. This determination becomes critical to assessing the contrast of proposed surface mining activities. Related to this is the second component, the cumulative visual impacts of surface mining activities in an area or region. Large surface mines, such as gold, coal and hard rock minerals are long-term operations, subject to incremental expansion as well as new operations adjacent to, or disconnected from, existing mining operations. In terms of visual impacts, and the need for visualization of those impacts, careful attention must be paid to these cumulative visual effects.

Data reliability, accuracy and currency issues

Surface mining results in the exposure (visual) of sub-surface conditions. This is expressed in rock layers, highwalls, talus slopes, and extreme visual changes in form, line, colour and texture. Although mining engineers use sophisticated technologies to map sub-surface geological strata, this data is rarely detailed enough to accurately predict the visual appearance of these conditions when exposed on the surface during and after mining. As a result, visualization of these conditions, which may not be revealed for weeks, months or years, is difficult and subject to some degree of conjecture. In contrast to the landfill situation, in mining engineering CAD or GIS digital data is sometimes incomplete, or of a degree of detail and resolution sufficient for general mine planning but not for developing highly accurate visualizations of concurrent and post-mining visual conditions. In the case of surface mining, operations and plans are constantly changing. Therefore, CAD and related data are likewise changing; data which is accurate this month may be outdated the next. Short-term changes in mine operations and plans have long-term visual effects which are generally hard to predict.

Political/social factors issues

There are two major factors in this category. First, various political jurisdictions may be involved in the regulation of surface mining activities on any one site. Federal, state and county jurisdictions are usually involved, with conflicting guidelines and regulations which the surface mining operators must try to appease. Second, in most cases the public will express some level of concern or sensitivity about the visual effects of proposed surface mining. Responding to this concern is one of the strongest aspects of visualization. 'A picture is worth a thousand words' has never been more true. Many of the environmental consequences of surface mining can be more clearly understood by the public (and the professionals) when seen in a properly developed visualization image. Visual simulations can be used as social survey instruments, identifying the visual resource issues of public concern and gauging public reaction and preferences for alternative mining plans. The public's 'sensitivity level' towards proposed visual changes is one of the major components of many visual resource analysis and management systems. Related to the operational issues described above, the future time periods illustrated must be carefully chosen, and must reflect changes in the landscape other than surface mining (for example, new roads, more people therefore more housing and commercial activity, growth and change in vegetation, etc.). Many of these changes are very difficult to predict and to visualize.

Visualization process issues

It is commonly assumed that a visual simulation must represent a high level of geometric accuracy in order to be useful and credible. In fact, research has shown this is often not the case (Watzek and Ellsworth 1994). This research determined that the scale of a proposed landscape change, such as a road or a pipeline, in some cases may vary as much as plus or minus 15 per cent before the scale inaccuracy is noticed by observers. As mentioned above, in surface mining, the design and planning process most often precludes highly accurate and reliable data for visualization. In many, if not most, planning studies at the EA or EIS level the project alternatives have not yet been designed in sufficient detail to support highly accurate visualizations. Humans are amazingly adept at 'filling in the blanks' in the visual display, which includes subtle or even gross inaccuracies in content, space, form, line, colour, texture and scale. This facility to understand the environment with minimal information means that landscape visualization can be successfully achieved with reasonable and defensible results without expending exorbitant amounts of time, energy, resources, and money in trying to achieve exact accuracy.

In the visual simulations of surface mining described above, design plans provided by the mining engineers were carefully studied and discussed, with multiple client and in-house reviews conducted with EALA staff. The proposed landscape changes were in the EIS design stage, therefore high levels of accuracy were not attainable or necessary. The purpose of these visual simulations was to communicate the intent of the mining plans to the government regulatory agencies, the mining company and interested citizens while recognizing that the plans were flexible and subject to discussion, review and change as necessary.

For visualizations that do not require three-dimensional or animated displays, sophisticated and affordable image-editing software and hardware is readily available for the development of highly realistic and credible visualizations (such as the surface mine visualizations seen above). This family of software works very well in those situations where highly accurate and reliable data is unavailable (as discussed above). Such software provides a very powerful array of image manipulation and enhancement features.

The starting point for effective and reliable image manipulation is appropriate base images. Cameras, either film or digital, are the basic tool in the field photographer's box. Either media works well, with film having the advantage of extremely high resolution but is relatively expensive per frame compared to digital which is also more flexible. The 'normal perspective' lens (e.g. 50 mm lens on a 35 mm camera body) is usually preferable over telephoto or wide angle to maintain proper perspective. Panoramic images can be achieved by digitally 'stitching' two or more normal perspective images together or with the use of a special panorama camera and lens.

Other useful equipment includes tripods (for stability but also for maintaining observation position), measuring tapes for determining camera-to-object distances and dimensions of elements in the landscape, GPS units for recording accurate locations of observation position, and helium balloons for determining accurate heights as well as extents of proposed landscape changes.

Careful and detailed records of images' location, weather and clouds, direction of view, aperture and shutter speeds, date and time of day, and other pertinent data are essential. All photo images should be captured at approximately the same time of day, preferably in the middle third of daylight hours. This will minimize the effects of long shadows which can be hard to match in a visual simulation. Plenty of photos should be taken from a variety of angles and observer positions (inferior, normal, superior) relative to the landscape undergoing change. Angles and positions should be systematically changed slightly with each successive shot to improve the chances of having just the right image when the computer work begins. All photos should be taken within a few days of one another, to minimize the effects of seasonal change in light and the associated effect on colour and texture in vegetation and other landscape features. Include people, cars, power lines, fence lines, and other known figures and features for scale (cows and other livestock work well in rural scenes as most people are familiar with their size). Use measuring tapes to confirm the dimensions of these elements as much as possible and keep meticulous notes.

The selection of important viewpoints, often called 'Key Observation Points' (KOPs) is crucial and includes careful attention to the angle of observation, number of viewers, length of time the project is in view, relative project size, season of use and light conditions. The most important viewpoints are often:

- from communities, road crossings, gathering places (e.g. shopping centres, churches, parks);
- typical views encountered in representative landscapes, such as the daily commute driving route of many people;
- points from which any special project or landscape features such as skylines, bridges, utility lines, etc. can be seen.

Conclusion

Photo-realistic computer visual simulations are a very valuable tool in the design and planning of surface mining or landfill operations. When used ethically and with care, they can result in representative, realistic, bias-free and credible representations of future conditions. They can be effective communication tools for explaining the operator's intentions, and indispensable analysis tools for assessing the visual impacts of operations on public and private lands.

This work is copyright © 2004 John C. Ellsworth (surface mining portions) and Abraham N. Medina (landfill portions), all rights reserved; printed here with permission.

THE PROVISION OF VISUALIZATION TOOLS FOR ENGAGING PUBLIC AND PROFESSIONAL AUDIENCES

David R. Miller, Jane G. Morrice and Alastor Coleby

Introduction

As the outcomes and consequences of landscape planning decisions are generally poorly shared amongst different stakeholders, there is a demand for tools that increase the understanding of landscape changes and provide techniques for supporting the planning, and negotiation, of changes in land use policy that will affect regional and local environments (Krause 2001; Appleton *et al.* 2001).

Such tools should enable public engagement in planning, which is a core element of the Aarhus Declaration on access to information, and public participation in decision making (European Union, 1998). In the UK it is recognized by public authorities and agencies, with Scottish Natural Heritage identifying a 'supporting programme of public engagement and awareness' (SNH 2003). Similarly, guidelines to the planning system in Scotland seek 'community involvement, dialogue and negotiation' as part of a process 'that respects the rights of the individual while acting in the interest of the wider community' (Scottish Executive 2003).

One definition of 'engagement' is provided in a report published by the UK Office of the Deputy Prime Minister (ODPM 2003). It states that 'engagement means entering into a deliberative process of dialogue with others, actively seeking and listening to their views and exchanging ideas, information and opinions, while being inclusive and sensitive to power imbalances'. The ODPM (2003) also notes that engagement should be undertaken not only when there is a dispute to be resolved, and that raising awareness and discussing topics with a wide audience can be undertaken over a period of time to develop a relationship between stakeholders in a geographic area, or associated with a particular theme. However, it should be recognized that the dialogue must be understood and must inform discussion among public and professionals alike.

Visualization tools can address some deficits in information provision (Bulkeley and Mol 2003), increase the effectiveness of decision making and potentially avoid the costly process of public enquiries. This chapter describes initiatives taken to support public engagement, and the communication of changes in the landscape to both professional and public audiences.

Landscape change

The consequences of change in the landscape vary in magnitude and time-scale. The high public profile of some prospective changes, such as those driven by targets to increase the amount of energy produced from renewable sources, is an example of how rapid change, perhaps for a limited duration (25 years), can arouse considerable interest. The importance of wind turbines in the landscape cannot only be attributed to their potential visual impacts, although this is one of the issues most often raised and argued over (http://www.communitypeople.net/interactive/distopics.asp).

Other landscape-related issues may have a lower public profile, but be of greater direct significance over the longer term. For example, the reform of the Common Agricultural Policy (CAP) (European Union 2003) provides support for rural development activities, and promotes production practices that are 'compatible with the maintenance and enhancement of the landscape'. There would appear to be the potential for large sums of money to be directed towards rural landscapes across the European Union, with a key requirement being consideration of environmental quality. Thus, over a long term, perhaps a period greater than 20 years, there is the opportunity of significant investment in the rural landscape.

Changes to the landscape are likely outcomes of the policies identified above (indeed there is an overlap, with CAP reform enabling investment in biomass under the rural development agenda). However, the time-scales over which discernible change may occur will be different, as will the nature of the changes. Wind turbines will introduce man-made structures taller than any current equivalents into a predominantly rural landscape, located on relatively few sites (but visible from considerable distances). In contrast, CAP reform may lead to an increase in apparently semi-natural features (e.g. woodlands) and maintenance of farm infra-structure (e.g. field boundaries using traditional materials) in very many locations across an area but, taken individually, features at any one location are unlikely to be highly visible. The potential changes in each example merit engagement with professional and public stakeholders, but some form of consultation is only likely with regard to wind turbines. In each case there could be a similar approach to the provision of materials to be used in the engagement with a public audience, including the use of computer-based landscape modelling to show and explain the extent of prospective changes.

Communications with members of the public

A recent example of engagement placed an emphasis on the two-way interactions between stakeholders in two woodlands purchased by a

Applications in energy, industry and infrastructure

a

b

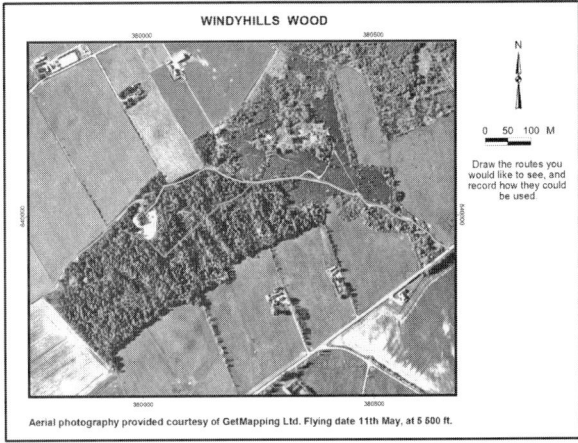

c

d

local community (The Woodhead and Windyhills Community Trust). An event was organized under the auspices of 'Planning for Real' (www.communityplanning.net/methods/method100.htm) to obtain local views on public access and use of the woodlands.

The main objective of the event was to obtain views and information from members of the community, but also to exchange opinions and raise awareness of issues associated with the use of the woodlands. Three different options were provided for gathering information on routes used through the woodlands, from which individuals could chose the media that attracted their attention or suited them the most.

Option 1 was a physical model of the area constructed by children of the local primary school (Figure 8.7(a)), on which people identified the pathways that were used through the woodlands, and the purposes to which they were put.

8.7 Stimuli used in obtaining local views on public access and use of the woodlands (a) physical model of woodlands site, with coloured tape showing preferred routes through the woodland and 'flags' indicating comments about particular locations; (b) aerial image of one woodland (Windyhills Wood), on which preferred routes were identified by participants; (c) ground-level view of the woodland, showing billboard models of trees and footpath from aerial imagery; (d) attendee at public event being introduced to the use of the computer model of the woodland

Option 2 comprised high-resolution aerial imagery of the woodlands, including the area surrounding the village, enabling paper-based (i.e. drafting of routes on acetate) or computer-based (i.e. digitizing of routes in ArcView), means for recording access routes and pathways through the woodland (Figure 8.7(b)).

Option 3 was an interactive computer visualization of the woodlands, through which people could navigate their own routes. Figure 8.7(c) shows a view of the model, and Figure 8.7(d) shows a participant being introduced to the use of the model and the navigation tools.

The basic input data comprised:

1. texture – high-resolution ortho-photographs (edited to ensure a colour balance across the images that would not negatively impact upon the visualizations);
2. topography – a DEM with a horizontal resolution of 5 m × 5 m;
3. feature models – three-dimensional models and billboards of individual features, of walls, buildings, trees and animals.

The ortho-photographs were draped across the DEMs to provide background textures and, with additional coverages from Ordnance Survey MasterMap (Ordnance Survey 2003), a basis for locating individual features in the landscape models (e.g. individual trees). Information on forest tree species was obtained from forest stock datasets. A set of billboards was created to represent a few examples of each species (e.g. Sitka spruce) of different heights. The sky has been created using a blue tone, into which clouds of different dimensions and densities have been introduced for background. The software used was ERDAS IMAGINE VGIS. The model was prepared with pre-set viewpoints, an option of pre-prepared walk-through routes, and a free choice of movement.

The availability of the alternative options for contributing comment on the preferred routes through the models appeared to attract a range of participants. Feedback suggests that participants felt obliged to use the physical model of the site. However, anecdotal responses suggest that interpretation of the aerial imagery provided an equally accessible media on which to record route information, with the added value of detail in the imagery to which people could relate, and triggered additional interest, and feedback (e.g. comments made on the level of waste visible, and experienced, in parts of the woodland).

The three-dimensional model proved to be of curiosity value, but also gave people the feeling that there was sufficient interest in 'their site' to spend time on the evaluation of options. For a minority of attendees the three-dimensional model provided a practical basis for their engagement in the identification of routes, via the recording of their walk-through and documenting comments at key viewpoints (e.g. the underfoot conditions, the popularity of the undulating path for mountain biking, and availability of distant views from particular locations).

Currently, the use of such models could still be argued to be contributing added value, and widening the range of materials and media by which communications can be undertaken. However, it is anticipated that such models will soon be considered one of the core means of engagement, particularly when combined within a virtual reality environment.

Communications between landscape professionals

A second example of the provision of materials for communication of the consequences of landscape planning decisions is directed at those with a professional interest. In this example some prototype quantitative measures of landscape views complement the use of three-dimensional computer models, with specific reference to the development of wind turbines. Provisional layouts have been considered, with a proposal submitted to the planning authority by the developer.

Standard photomontage approaches have been used in the professional evaluation of the proposal, together with wire-frame models and Zones of Visual Influence (ZVIs). Such information comprises the core requirements of planning authorities in the UK for evaluation of the visual impact of changes in land use, following the guidelines published by the Landscape Institute and the Institute of Environmental Management Assessment (2002).

In a study of the potential future changes in land use for the same area, visualizations using model features and texture backdrops have also been created from a model developed in ERDAS IMAGINE VGIS. The model provides for a walk-through of the landscape, through the woodland and to a vantage point close to one part of the development. Figure 8.8(a) shows a perspective view of Clashindarroch Forest in north-east Scotland on which a wind turbine development is proposed. Figure 8.8(b) shows a close-up view, with the model features of the woodland in the foreground, along with a mast for a mobile telephone network. Such imagery has provided the flexibility of illustrating the potential views of the turbines from different locations, and with alternative layouts.

However, the evaluation of such developments includes other considerations, such as that of the character of the landscape. Miller *et al.* (2004) have produced a model of the potential for wind turbine development in northern Scotland for Scottish Natural Heritage (SNH). This model has been developed using largely quantitative measures of the terrain and natural features, such as land cover based on the principles proposed by Bell (1998).

In the derivation of landscape character sensitivity to turbine development, a similar approach to that of the ZVI mapping has been adopted, but which provides inputs to a model of perceived visual nat-

Applications

8.8 Views of Clashindarroch Forest: (a) perspective view looking north-east, with a provisional layout of wind turbines; (b) close-up view of wind turbines and the intervening trees

uralness of the landscape. The method is based upon the calculation of the number of cells that are visible from every other cell to represent the extent of the visible area, as initially proposed by Tandy (1967) in relation to the description of landscapes. Benedikt (1979) then considered the interpretation of the relationship between the viewer and their surroundings in terms of measures of perimeter and area of levels of visibility, in an architectural context. Steinitz (1990) applied similar approaches to the evaluation of the content of views from tourist routes, with early trials of computer-based modelling with GIS. O'Sullivan and Turner (2001) have subsequently applied a variation on the isovist analysis to the derivation of measures of the landscape which may be related to human perceptions of its character.

The extent of perceived naturalness within the view of the observer was mapped by using a viewshed-based analysis of individual land cover types (Miller 2001), and comparing the proportions of semi-natural land cover in the view with those of types of human origin, for example, the relative amounts of semi-natural montane or moorland vegetation compared with cropland or settlement. The categorization of land cover classes was into 'mainly natural' or 'mainly man-made', and the output was the percentage of the view from each 10 m × 10 m cell from which 'mainly' natural land cover was calculated to be visible. The measures of naturalness have been refined by processing the data for different distance bands so that distance decay can be considered within the model of the view content.

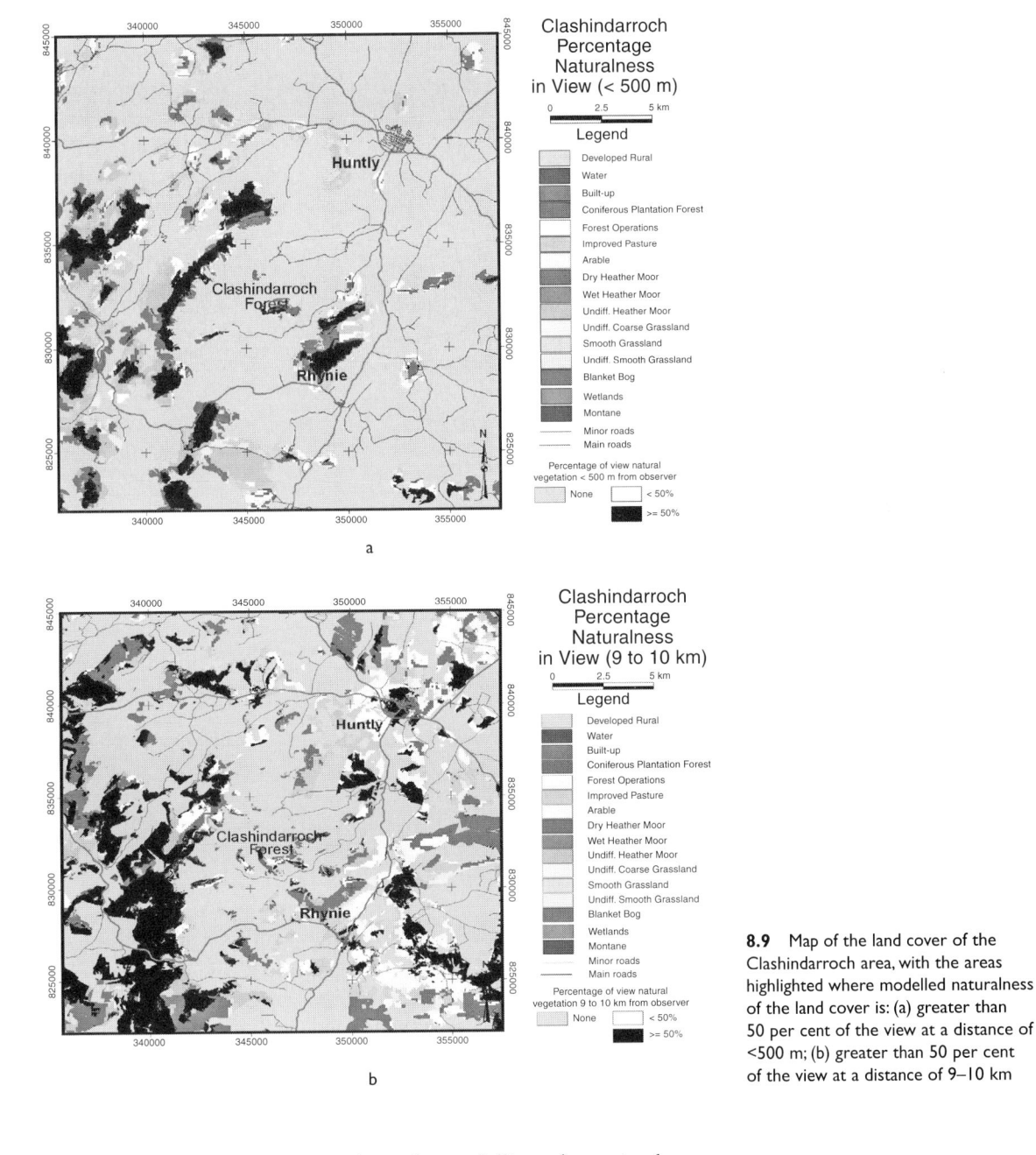

Applications in energy, industry and infrastructure

8.9 Map of the land cover of the Clashindarroch area, with the areas highlighted where modelled naturalness of the land cover is: (a) greater than 50 per cent of the view at a distance of <500 m; (b) greater than 50 per cent of the view at a distance of 9–10 km

Figure 8.9(a) shows an output from the modelling of perceived naturalness of the land cover in the viewshed, for a viewing distance of up to 500 m, and Figure 8.9(b) shows the output for the distance 9 km to 10 km. The results show the distribution of the areas from which the land cover visible is categorized as natural, with the greatest extents being in the south-west of the area, where the moorland is dominant. However,

Applications

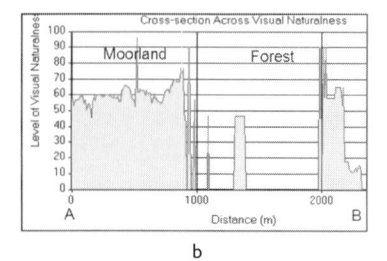

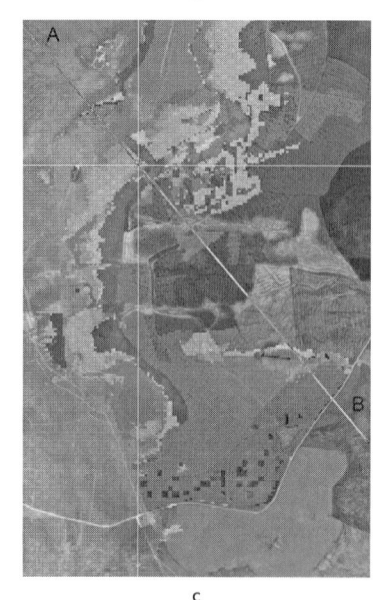

8.10 Interpretation of model of naturalness of the landscape in view with respect to turbine locations. Transect A–B shows the level of visual naturalness, decreasing from west to east, with a drop to zero at the forest edge: (a) perspective view of model of naturalness draped across orthophotograph, with cursor querying location (also shown in (c)); (b) transect; (c) orthophotograph with model of visual naturalness overlaid (line of transect A–B shown)

the land to the east and north also provides vantage points that offer views of natural land cover types at distances of between 9 km and 10 km. The presence of the turbines will have lower effects upon the perceived naturalness at the greater distance from the observer and it is the land that is close to the area in which the turbines will be located which will experience the greatest impact of the development of the turbines.

Figure 8.10 shows the area in a perspective view, with the naturalness surface draped across the terrain model, and an example layout of the turbines placed into the model. In this view one has a more intuitive approach to interpreting the areas from which turbines will contribute to reducing the high levels of naturalness in the view, and thus impact on the landscape character sensitivity to turbine development.

By exploring the representation of derived data in three-dimensions one can add to the portfolio of options for communicating the effects of landscape change on the visual landscape. Such an approach to the presentation of derived data on landscape character would be of greatest relevance to the expert, and probably offer little to the general public's understanding of the change.

Conclusion

With changes in the landscape being driven by a number of national and international policies, significant sums of money are likely to be invested in the rural environment, which will lead to changes in both the short and long term. Whether such changes are built structures, or changes in vegetation, they could prove to be considerable in their impact. The role for tools that can be used to explore scenarios of change will become increasingly significant as evaluations of potential changes are required by both policy-makers and field officers.

In each case, recognition of the importance of retaining public support for investment in the countryside will remain high, and initiatives to encourage engagement, in all of its forms, are likely to be promoted.

Acknowledgements

The authors would like to thank the Scottish Environment and Rural Affairs Department, the European Commission (through contract QLK5 – CT – 2002 – 01017) and Scottish Natural Heritage for financial support for this work.

THE VISUALIZATION OF WINDFARMS
John F. Benson

Introduction

Global warming and climate change have stimulated a wide range of responses around the world, significant among which are moves to increase the proportion of electricity generated from renewable sources. The technology that has reached the market in large quantity is wind turbines and windfarms. Deployment of this technology has been controversial (e.g. Kahn 2000) and the dominant and recurrent causes of controversy have been the landscape and visual effects and impacts of the structures (Pasqualetti 2001; Pasqualetti *et al.* 2002). In part, this is not surprising; windfarms, like overhead power lines and other components of society's energy infrastructure, are often located in rural, remote or wilderness areas, and in the case of the visual effects of windfarms, they are impossible to mitigate.

Wind turbines and windfarms can change the character of the landscape and can also have visual effects. Here, visual effects are taken to be a subset of landscape effects, in accordance with current guidance and practice in the UK (LI-IEMA 2002; Hankinson 1999), and the focus here is on such visual effects. Harnessing energy from the wind is an old technology and the windmills of Holland and elsewhere evoke nostalgia and even some of our favourite literary imagery (De Cervantes 1605). In contrast, modern wind turbines are industrial structures, albeit with a partly-benign or positive environmental purpose, that evoke strong feelings (Krohn and Damborg 1998). Thayer argues that opposition to wind turbines can be explained partly by what he terms landscape 'guilt': technophobia is a source of environmental guilt, which is then commonly manifested in the actual, physical landscape by many examples that involve deception, concealment and camouflage (Thayer 1994). These issues are controversial.

It is probably not controversial to state that planning, participation and the decision-making process all need to be based on, among other things, accurate and realistic visualizations. Accuracy and realism in visualization are not the same things, and these are only precursors to the assessment of significance; these key issues are explored in this section, based on a recent study in Scotland (Benson *et al.* 2002).

General findings

The development process for many wind turbines and windfarms requires formal Environmental Impact Assessment (EIA) and the incorporation of the results into an Environmental Statement (ES). There appears to be a great deal of variation in the way that Visual Impact Assessment (VIA) is dealt with in EIA. Also, there is remarkably little research conducted on the accuracy of the predictions that lie at the heart of EIA (Wood *et al.* 2000).

We reviewed eight windfarms built in Scotland during 1995–2001. The windfarms studied were Beinn An Tuirc (2001), Beinn Ghlas (1999), Deucheran Hill (2001), Dun Law (2000), Hagshaw Hill (1995), Hare Hill (2000), Novar (1997) and Windy Standard (1996) (the date is the date of construction or commissioning; the respective ES was prepared between 1–4 years before construction). Sizes ranged from 53.5–85.5 m to blade tip. The basic procedure was to compare, at the specific viewpoints used in the ES, the as-built visibility with the predicted and simulated visibility. At the same time, we sought explanation for the probable or possible causes of any differences we found.

Most parties in the debate welcome the use of guidelines, but many guidelines (for ZVI, VIA and other factors) for assessing windfarm developments are based on first generation turbines and these need to be revised for second and third generation machines (see below). There is some research and a wide and diverse range of guidance and opinion on the detailed issues surrounding ZVI, distance, visibility and the significance of visual effects. This diversity is partly explained by the complexity and the subjectivity of the issues. Equally important is the desire by one set of windfarm interests to minimize the political, professional and public perceptions of the visual (and landscape) effects of windfarms (the wind energy 'industry', broadly defined) and an opposing desire by another set of interests to maximize these perceptions (the critics and opposition groups, broadly defined). As a simple example, the pro-lobby might say 'only visible in clear conditions' whilst the anti-lobby might say 'always visible in clear conditions'; both statements are, of course, accurate, true and in a sense identical, but such subtle if trivial bias is widespread in practice. Visualization and then VIA are therefore highly politicized activities.

Accuracy in Environmental Impact Assessment

Zones of Visual Influence (ZVI) were always calculated and inserted into the ESs examined (although the term Zone of Theoretical Visibility might be preferred; Hankinson 1999). There was considerable variation, often unexplained, in the technical specifications for the ZVIs, including the datasets used, the method of interpolation, the outer limit distance selected and so on. These ZVIs were never wholly accurate, nor was it ever claimed that they were wholly accurate (see also Bishop 2003). We found very little research that has attempted to quantify that inaccuracy (Wood 1999, 2000, are exceptions), especially in any useful formula that could be used in practice.

Every case studied used wireframes and photomontage. The number produced, and the viewpoints selected, appeared to result from negotiation between the parties on a case-by-case basis. Some viewpoints were selected to predict the appearance of the windfarm from sensitive locations. Some were selected to predict the appearance from representative locations. Others were chosen to test the accuracy (or otherwise) of the ZVI – they produced not-visible visualizations! Three quite separate purposes for the use of photomontage are operating here. This is an important distinction in the use of visualizations in practice.

If visualizations are being used for what are effectively three separate purposes:

1. to assess the potential significance of effects at key viewpoints,
2. to provide a representative selection of visual effects (essentially this is visual survey and not assessment), and
3. to test the ZVI,

then these three purposes should be distinguished. Although mixing these purposes might appear harmless, it can result in an ES that contains potentially and superficially very misleading information. For example, statements to the effect that 'the windfarm would only be visible from three of the fifteen viewpoints assessed' can be inserted freely and truthfully, without acknowledging that twelve of these may have been selected specifically to show just such a non-visible and therefore non-significant effect.

While wireframes could be thought of as working tools, and photomontages as presentation tools, we think both can serve both purposes. Wireframes can be preferred to photomontage because they reduce the risk of implying a false realism (see below). In general we found that both tools were accurate in the sense that they placed the turbines in the correct locations and at the correct height in relation to the topography (notwithstanding some simple errors, and frequent post-EIA design changes).

Realism in visualization and photomontage

Photomontages need to be realistic as well as accurate. Perkins (1992) asks what influences 'perceived realism'? He points out that realism may be affected by the context or content of the image portrayed. A technically accurate and precise photomontage that placed Edinburgh Castle on Cairngorm Mountain will not be perceived as realistic for obvious contextual reasons. Although less extreme, a proposed windfarm placed in a remote landscape may be perceived by a viewer as containing an element of incongruity and inappropriateness that will affect their evaluation of the visualization.

Applications in energy, industry and infrastructure

Our general finding, based on 113 subjective but systematic evaluations at 70 viewpoints, was that there was a recurrent tendency to underestimate the magnitude of visibility in the ES descriptions compared to assessments we made on site. This underestimation had two components. First, we frequently judged that the magnitude of the effect on site was larger than the verbal declaration made or predicted in the ES. Second, the windfarm often looked nearer, more visible and more conspicuous than the photomontage predicted (this is a frequent but unquantified and unverified comment made by professionals with whom we have discussed this issue). The two are probably related; if the professional assessor uses her own photomontage to predict magnitude (and then to assess significance), the photomontage probably causes the underestimation of magnitude and, of course, then results in an underestimation of significance.

As Figure 8.11 illustrates, the main reason for this appears to be the widespread practice of devising, sizing and printing photomontages that have to be viewed at unnaturally short distances (a further factor may be movement, discussed below). For example, in the eight ESs reviewed, the recommended viewing distances for the photomontages (where stated) ranged from 17–24 cm. Our judgment is that this configuration is a strain on the eyes, is difficult or impossible to use and fails to capture any semblance of realism. Because most viewers will, in practice, observe these images from longer distances, a subtle but powerful under-representation of the visual effect is introduced.

8.11 Each picture (or photomontage) is identical in every respect, except for the printed size. Which delivers the most accurate visualization of the windfarm? Which, if any, delivers a realistic visualization for you of the windfarm as you imagine you might perceive it on site? It is suggested that the largest image needs to be enlarged at least fivefold to begin to approach a realistic rendition

A typical, comfortable viewing distance for reading A4 pages is 30–40 cm, and a typical, comfortable viewing distance for larger images at either A4 or A3 held at arm's length is 50–60 cm. We therefore conclude and recommend that what is comfortable and natural for the viewer should dictate the technical detail and not vice versa. This means that visualizations should be designed for typical viewing distances of 30–50 cm and that most visualizations should be correspondingly larger. A full image size of A4 or even A3 for a single frame picture, giving an image height of approximately 20 cm is therefore to be preferred, rather than the common use of images with a height of approximately 10 cm. Occasionally, panoramas were produced using wide-angle lenses, which causes misleading distortions; splicing standard photographs is to be preferred, and the focal length of the lens and camera format used for photographs (and derived visualizations) should always be stated. Use of a 50 mm lens in a 35 mm format is recommended, or equivalent combinations in other formats.

Our main conclusion on the causes of this widespread effect of underestimation – small photomontages – appears at first sight to be so obvious and simple that it is remarkable that what seems to be current practice, at least in the UK, is allowed to continue relatively unremarked, unchallenged and unsolved. The effect was flagged by Stevenson and Griffiths (1994) but appears not to have been translated into improved practice.

Dynamic technologies

The technology of electricity generation and the technology of visualization are dynamic processes, while wind turbines are also dynamic structures.

The first contemporary commercial turbines erected in the UK were at Delabole in Cornwall in 1991. These are rated at 400 kW, stand 32 m to the hub and 49.4 m to the blade tip. In 2003, such machines are no longer manufactured and second and third generation structures, rated from 600 kW up to 2 MW, and up to 120 m to blade tip, are widespread. Because it is the blade-swept area that determines the power output, the commercial and economic trends towards increases in size are powerful. The effect is to increase scale effects in relation to landscape character change and to increase the distances over which visibility effects occur, increasing further the need for accurate and realistic visualization.

Back in 1991, common desk-based PCs in offices might take days to calculate a ZVI (if they were capable of it at all) and production of a photomontage would involve relatively high costs that limited their use to a small number for each development proposal. Twelve years later, the equivalent PC probably has ten times the power, a wider range of dedicated software for windfarm planning and visualization is available (e.g. WindPRO, WindFarmer, WindFarm), and society's expectations for simulations and visualizations have increased, including (at least on an experimental or leading-edge basis), the use of video, virtual reality and other tools. None of the case examples we reviewed used these latter tools. In passing, one might note the paucity of good research to compare these alternative visualization tools; it is not at all obvious that the more advanced tools necessarily improve the accuracy or realism of visualizations (although a unique feature of wind turbines – movement – can be simulated), and they will themselves be fraught with subtle and not-so-subtle dangers of bias.

It was found that the movement of the blades, in all cases where this was visible, increases the visual effect of the turbines because it tends to draw the eye (Bishop 2002, has quantified this effect using digital simulations). Movement could be detected with clarity at distances up to 15 km in clear conditions or conditions of strong contrast between the rotors and the sky, but only if the observer was specifically looking for the windfarm. On occasions, movement was not visible at 6 km in weak contrast. At a distance of more than about 12 km blade movement can become hardly perceptible and blade movement is judged to be perceptible to the casual observer at up to approximately 10 km. Movement was more perceptible when back-dropped against dark vegetation compared to grey sky. Photomontage cannot capture such movement and the issue was treated weakly or hardly at all in the case studies.

The current dynamics of deployment also mean that cumulative or additive effects could arise from many small (insignificant) changes, several large (significant) changes or interactions between wind energy and other types of development. Cumulative effects on visibility can arise from intervisibility, simultaneous visibility, visual coalescence and sequential visibility. Two turbines or two windfarm sites might be *intervisible*, one from the other. Also, although they may not be intervisible, they might be *simultaneously visible* from a viewpoint or a travel route. When *simultaneous visibility* occurs, there is also the possibility of *visual coalescence*. Whether coalescence occurs depends on the viewer's location and the locations of the separate windfarms and their relative heights and distances from the viewer. Finally, they might be *sequentially visible* as the observer moves through the landscape. Such sequential visibility will be more pronounced if the distance(s) between windfarms are short. A key factor is the time between sightings, so that a car driver will experience the sequential effect over longer distances than a walker. Frequent or repeated sequential visibility can then lead to the perception of a wind energy landscape, where the wind turbines become the defining characteristic of that landscape. Cumulation was not yet a significant issue in the cases examined, but the visualization of cumulation can and should be addressed; video and other dynamic tools hold promise in this respect.

From our study in Scotland, we also formed the impression that there was a conspicuous lag between the latest visualization tools and the examples of practice that we saw demonstrated. Further, in the case of research and even practice-based research on issues such as perception and significance, this had failed to penetrate quickly into EIA and VIA practice. These are issues that need to be addressed.

Perception and significance

Accurate and realistic visualization is, of course, only the starting point for good impact assessment. The magnitude or size of wind turbine elements, and the distance between them and the viewer, are basic physical measures that affect visibility. The key issue, however, is human perception of visual effects, and that is not simply a function of size and distance. The influences on apparent magnitude are many and varied, including the size and scale of the development, proportional visibility, lighting, movement and orientation, distance, colour and contrast, skylining and back-clothing, elevation of windfarm and human receptor, and colour and design. Within each of these factors, some effects tend to increase apparent magnitude and some tend to reduce it. This is why the specific environmental conditions at or affecting each viewpoint should be stated and analysed, including factors such as season and weather, air clarity, movement, orientation to prevailing winds and visual cues, as well as the detailed design and layout of the windfarm. These are discussed more fully in Benson *et al.* (2002), where a new conceptual model and schema for assessing visual effects is presented (Figure 8.12). A key conclusion is that photomontage (and probably other, more sophisticated, visualization tools) can probably never capture or reflect the complexity lying behind apparent magnitude and the perception of visual effects; realism may in this sense be forever elusive.

All visualization work, including attention to accuracy and realism, is likely to be of limited use or unpredictable effect if society cannot address the critical issue of significance. Of all the issues surrounding VIA, significance is the most subjective and intractable, while prediction and then evaluation of significance are at the heart of EIA. Remarkably, perhaps, significance is little researched in relation to visual impacts. Exceptions are Bishop (2002) (and see Shang and Bishop, 2000) and Stamps (1997), who offers a detailed review of the issue (including the related issues of design guidance and design review) and a theoretical and methodological model for assessment based on a statistical analysis of human preference ratings for before and after scenes. However, his focus, and his case studies, are based on urban design issues in California. Until robust consensus on significance can be claimed with confidence, best practice requires that the bases for all judgments made are clear and explicit on a case-by-case basis.

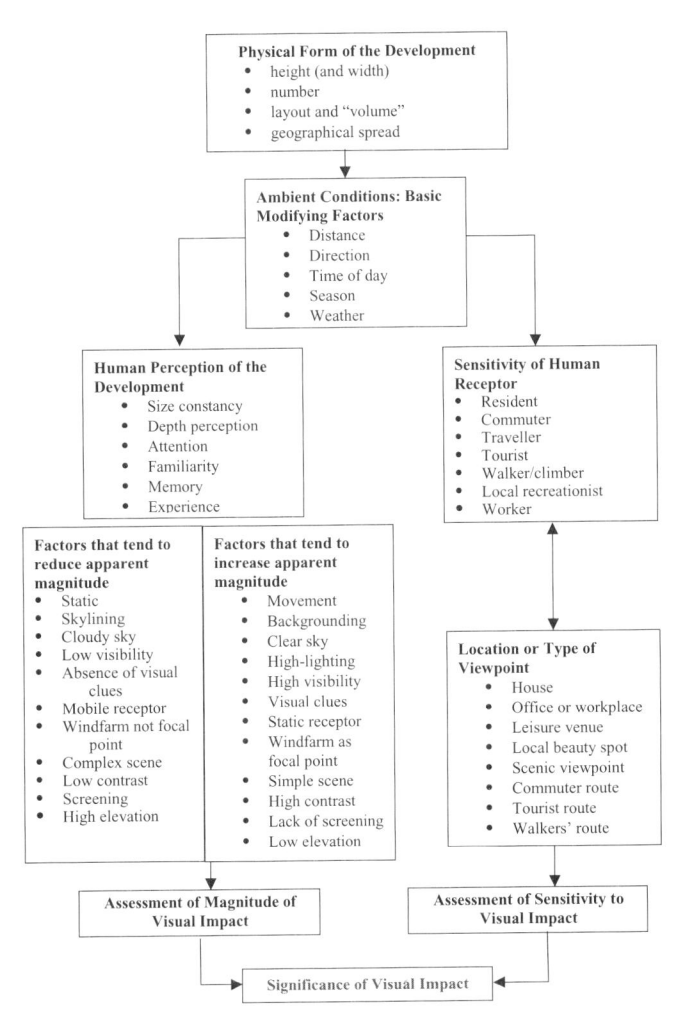

Applications in energy, industry and infrastructure

8.12 Conceptual model for Visual Impact Assessment (see Benson *et al.* (2002) for further details)

Conclusions

The increasing development pressures for windfarms require that visualization and VIA are approached in a comprehensive, explicit and systematic way and that the inherent complexity, controversy and uncertainty are addressed. A photomontage can imply a degree of realism that may not be robust, and can seduce even a critical viewer into investing more faith in that realism than may be warranted. Accurate and realistic visualizations are only parts of a complex analytical and political process.

Acknowledgement

The findings reported here are drawn from a study commissioned and published by Scottish Natural Heritage, but the interpretations and conclusions are the author's and should not be taken as representing SNH opinion or policy.

CHAPTER 9

APPLICATIONS IN THE URBAN LANDSCAPE

INTRODUCTION
Eckart Lange

Demographers predict that by the year 2020 nearly 60 per cent of the world's population will inhabit urban areas (Hall and Pfeiffer 2000; UNCHS 1996). Consequently, it is not surprising, that urban green space is increasingly being perceived as an important factor for sustainable development and human well-being. For a long time green spaces were just viewed as remnants or by-products of urban development (Selle 1999). Now, the British Department for Transport, Local Government and the Regions (DTLR 2001), for example, claims that strategic planning of parks and urban green spaces must be seen as paralleling and augmenting strategies for housing, community development, safety and economic regeneration.

Nowadays, attractive urban parks, woodlands, green corridors and other types of green spaces can be found in many cities around the world. However, urban green space is often a legacy of earlier decades. Already in the mid-nineteenth century the importance of urban green space was recognized. Perhaps the most famous examples for the strategic planning and design of urban green space are Central Park in New York (1858–1876) and the so-called Emerald Necklace in Boston (since 1878), both by the landscape architect Frederick Law Olmsted (Rybczynski 1999). Also, the town planner Camillo Sitte acknowledged and underlined the importance of urban green space. Sitte (1922) made a distinction between sanitary and decorative green. Besides fulfilling important health (e.g. de Hollander and Staatsen 2003), amenity

(Gälzer 2001) and social functions (e.g. Duffy 2003), urban green spaces function, in ecological terms, as the lungs and water filters of our cities and can provide important refuge and highly complex habitats for flora and fauna. In addition to assessing the landscape quality and recreation value of urban green space, Lange and Hehl-Lange (see the section 'Future scenarios of peri-urban green space', p. 195) are looking into the perceived ecological value of alternative future scenarios using three-dimensional visualizations.

Furthermore, green space and urban open space have a considerable economical importance. Several studies have shown that the economic value of a property depends significantly on its relative proximity to green space, water and open space (e.g. Conner *et al.* 1973). Also, it has been proven that the quality of a view is a critical factor (Lange and Schaeffer 2001; Bishop *et al.* 2004). In this context Danahy (see 'Negotiating public view protection and high density in urban design', p. 203) shows examples of visualization in public view protection and urban development. Zooming into great detail, Laing *et al.* (see 'Combining visualization with choice experimentation in the built environment', p. 212) investigate experimental physical changes in an urban streetscape environment.

Due to the continuous process of urbanization, urban green space is permanently under high pressure for development. On the one hand there is a strong trend of urban sprawl, using up highly productive agricultural areas. On the other hand, the often proclaimed infill development that reduces urban sprawl can also cause ecological consequences such as alteration of the micro-climate, soil-compaction, increased run-off or, ultimately, complete destruction of green space or ruderal habitats (e.g. Finke 2003).

In the presented examples from Aberdeen, Toronto, Ottawa and Zürich visualization is used as the key communication tool. It is employed to express visual and spatial ideas to stakeholders, focus groups and other interested citizens or expert groups. The purpose of the visualization is to provide the foundation for the ways in which people interpret and react to the visual experience of the landscape and to specific characteristics of green space including vegetation and built structures. The visualizations are functioning as an instrumental basis for the assessment of landscape changes and the study of alternative futures ranging from the peri-urban fringe in Zürich, to urban parks in Toronto, to the symbolic core of Canada's capital Ottawa and to a highly sealed streetscape environment in Aberdeen.

FUTURE SCENARIOS OF PERI-URBAN GREEN SPACE

Eckart Lange and Sigrid Hehl-Lange

Introduction to the Käferberg case study site

For the citizens of Zürich the provision of green space and public transport are the two most important elements contributing to the quality of life (Fachstelle für Stadtentwicklung 1999 and 2003). While public transport scores the highest in terms of degree of satisfaction, the people of Zürich think that their green space has potential for improvement. Out of a total of eleven criteria it only achieves the sixth, i.e. a middle rank, in satisfaction.

In the presented example, the Käferberg recreation area is used as the case study site. Situated on hilly terrain between the campus of the Swiss Federal Institute of Technology and the Käferberg forest it provides good views towards the city with the Alps in the background (Figure 9.1) and the valley of the Limmat River. The Käferberg case study represents an open space, where ecological functions and agricultural production functions are confronted with extremely high visitor pressures. Within the sequential variation of the types of green spaces, it lies between the predominately formal green spaces in city centres and the more natural looking green spaces such as forests and farmland that can typically be found further away.

9.1 Virtual landscape model of Zürich with the Alps in the background. The Käferberg green space is in the foreground between the Käferberg forest and the campus of ETH Hönggerberg

Because of the easy accessibility by public transport, the availability of parking facilities and the high quality of the green space, the Käferberg case study site can be classified as being of city-wide importance.

In the Käferberg open space between 1961 and 1973 the first phase of the Campus of ETH Hönggerberg was built (Huber 1990). As a consequence of the beginning of the construction activities a process of transformation from rural to urban started that culminated in the new chemistry building of ETH.

Objectives

The Käferberg case study in Zürich concentrates on the public perception of aesthetical, recreational and ecological values. By using GIS-based visualizations the characteristic qualities of the Käferberg open space are reflected in three dimensions and are compared with the potential impact of alternative future changes. To get an overview of the opinion of the public, i.e. laypersons as well as expert planners, regarding the evaluation of landscape changes as presented by three-dimensional visualizations, a survey is conducted, both in traditional paper format and via the Internet. It is hypothesized that there is a great potential for improvement of the current landscape situation with regard to public perceptions of aesthetical, recreational and ecological values.

Visualization methodology

In the Käferberg case study static and dynamic representations (Appleyard 1977; McKechnie 1977) of urban green spaces are employed. The visualizations are used to conduct public surveys and to assist in participatory planning.

The digital models with various levels of ground resolution and the visualizations are derived using high quality three-dimensional data from GIS and CAD, as well as interpreted two-dimensional data, site survey data and site photographs. The goal for the models is to provide optimal resolution and detail while allowing the digital model to be computationally efficient.

For the visualizations, Polytrim software from the Centre for Landscape Research at the University of Toronto (Danahy and Hoinkes 1995) was used. It is a visualization software that has a strong link to GIS and CAD software (Hoinkes and Lange 1995).

The visual representation of the landscape at the Käferberg consists of a digital terrain model, an orthophoto and three-dimensional elements such as houses, trees and farm animals. For a precise representation of the existing vegetation, tree and shrub species were surveyed, mapped and photographed in the field. From these photographs, a texture library was established.

Scenarios

Throughout history, people have tried to make today's decisions by understanding the possible consequences for tomorrow. Prognoses and forecasts aim to predict a specific state of the future (e.g. Stiens 1993; Stremlow *et al.* 2003). However, no one can tell the future with certainty. In contrast, scenarios provide an array of possible or even unlikely future changes. Translating scenarios into visual representations has become increasingly popular in recent years. Examples include hand-drawn sketches used by Heißenhuber *et al.* (2000) or digital photomontages as used by Tress and Tress (2003). In these examples, bird's eye views are shown.

In the presented case, scenarios provide synthetic three-dimensional images of a real landscape as a visual outline of a spectrum of alternative futures for the Käferberg.

In the Käferberg case study the scenarios are developed around certain identified themes and driving forces. In this new approach to green space planning the process of scenario development involved the participation

9.2 (top left)   Overview: Status quo

9.3 (top middle)   Agriculture scenario

9.4 (top right)   Recreation scenario

9.5 (bottom left)   Nature conservation scenario

9.6 (bottom right)   Wind turbines scenario

(Images reproduced in the colour plate section)

of experts, landowners and land-users. They were asked to express their ideas using two-dimensional plans.

Based on the actual dominating land-uses and the various plans and ideas for change, scenarios were developed which reflected emphasis on the status quo, agriculture, recreation, nature conservation and wind turbines (see Figures 9.2 to 9.6). These were created as three-dimensional models allowing visualization from any viewpoint.

The scenarios differ from each other in location, quantity and quality of the vegetation structure (hedges, single trees, fruit trees, orchard, forest) and the presence or absence of pasturing livestock (cows, sheep or horses are added in selected scenes) and wind turbines. The existing and newly erected chemistry buildings of the ETH are present in all scenarios. Depending on the viewpoint and the viewing direction as well as on the placement of new vegetation elements in a scenario, the new building can be completely visible or completely hidden. Altogether, one overview scene and eight viewpoints on the ground, i.e. from the viewpoint of a pedestrian, are chosen to show the effects of the five scenarios.

Survey

In order to test responses to the different visualized scenarios a survey was conducted. The survey was prepared in paper format and as an Internet version online. For interested local citizens, stakeholders and special interest groups a complete printed version of the test set with all eight viewpoints and a total of 49 images was provided in a face-to-face situation or as a mail survey. The test set used in the paper-based survey was randomized for each test person.

In the survey the respondents were asked to indicate their ratings for landscape preference ('How do you like the above landscape?'), recreation preference ('Would you enjoy walking in the above landscape?') and nature conservation ('What value do you think the above landscape has in terms of nature conservation?') by rating the colour images using a five point scale.

The survey was also placed on the Internet. In order to keep the survey short and to reduce the time needed to do the test (important for modem, dial-up) the online version consisted of 20 images only. This reduced test set was randomly selected from a database of the overall 49 images each time a user clicked on the link for the survey.

To attract test persons, the survey was announced in major Swiss newspapers, local newspapers and through Internet platforms. Together, the paper-based version and the Internet-based version of the survey gathered more than 400 responses.

Results

All 49 images used in the test are ranked based on their means on a continuous scale ranging from 1 (very low value) to 5 (very high value). In the paper-based survey the highest ratings for landscape preference, recreation preference and nature conservation were given to an image (Figure 9.7) that shows the scenario nature conservation with cows. In the Internet-based survey the highest rating for landscape preference was also given to the same image.

The least preferred scene for landscape preference, recreation preference and for nature conservation was an image showing the status quo with the campus of ETH Hönggerberg and the new extension and hardly any structural elements in the foreground.

Comparing the Internet and the paper-based surveys, the ratings for the three aspects landscape preference, recreation preference and nature conservation do not differ very much. However, one notable exception is the rating for nature conservation in the web-based survey. There, the values given are consistently lower than the ratings for the other dimensions by about 0.25. A possible reason could be that ecological information might be difficult to communicate with images only and additional textual explanation is required. Furthermore, it might have to do with the composition of the sample. The Internet survey has a high proportion of environmental experts.

There is an overall tendency in all three questions for laypersons to give slightly higher ratings than environmental experts (comprising, for example, biologists, landscape architects and geographers) or experts for the built environment (e.g. architects and civil engineers). This is especially the case for those images in the higher to the middle range of scores.

On the other hand, there is also a difference in the rating pattern within the two expert groups. There are some images where the built environment experts provide quite different ratings in comparison to the environmental experts.

Looking at local versus non-local people for question 1 (landscape preference) there is also a distinctly different rating pattern. The locals

	Ranking	Mean
	Web/paper	Web/paper
Landscape preference	1/1	4.04/4.02
Recreation preference	2/1	3.99/4.04
Nature conservation	2/1.5	3.72/4.00

9.7 Nature conservation scenario, with cows

Applications

consistently value their green space higher in terms of landscape preference than the non-locals. This did not occur, however, for either question 2 (recreation preference) or 3 (nature conservation). Obviously, the test persons are able to differentiate between the three questions.

In order to test the influence of farm animals on landscape preference, recreation preference and perceived nature conservation value, eight pairs of images that were each identical except for the presence or absence of farm animals (cows, sheep and horses) were included in the test. Compared to scenes without livestock, for the most part there is a considerable positive effect when livestock are present in the scene.

Current trends in agricultural policy in Switzerland will tend to increase the presence of livestock in the landscape. Among conventional farmers subsidies of the increasingly popular ROEL programme (Swiss Federal Office for Agriculture 2003) help to provide regular outdoor exercise for livestock. More than 60 per cent of farmers are participating in that programme and our results suggest that this will be popular with the public.

In scenes that already achieve high scores the influence of animals in the scene is rather marginal. In scenes that achieve lower scores the presence of farm animals can contribute to an increase in the ratings. Horses, in particular, influence the ratings very positively. For all three questions in the paper survey the image with horses scores about 0.5 higher than the same image without horses (see Figures 9.8 and 9.9).

	Ranking	Mean
	Web/paper	Web/paper
Landscape preference	26/31	3.18/2.96
Recreation preference	26/30	3.21/3.00
Nature conservation	27/36.5	2.94/2.85

9.8 Hönggerberg: status quo

	Ranking	Mean
	Web/paper	Web/paper
Landscape preference	19/16	3.45/3.48
Recreation preference	11/16	3.54/3.51
Nature conservation	12/21	3.25/3.29

9.9 Status quo with horses

Table 9.1 Overall ratings for the five scenarios

Scenario	Landscape preference		Recreation preference		Nature conservation	
	Paper	Internet	Paper	Internet	Paper	Internet
'Nature conservation'	3.40	3.45	3.40	3.40	3.45	3.20
'Recreation'	3.11	3.17	3.12	3.15	3.15	2.93
'Agriculture'	3.01	3.06	3.03	3.06	3.03	2.82
'Wind turbines'	2.61	2.82	2.72	2.88	2.79	2.63
'Status quo'	2.62	2.85	2.68	2.85	2.72	2.63

Implications for landscape planning

Each image used in the survey belongs to one of the five scenarios. From the overall score of all images of a specific scenario, the rating of the scenario can be determined. Table 9.1 shows the mean values calculated on the basis of only those viewpoints where all scenarios are visible. Those viewpoints from which the wind turbines are not visible are omitted in the analysis of the scenarios. Also, the images showing farm animals were not used in the analysis, because not all farm animals were present in each of the viewpoints that were used.

All scenarios are rated at least equal to or higher than the status quo: sometimes considerably higher. This shows that there is great potential for improving the existing landscape. Results indicate a very strong preference for the scenario nature conservation. It is followed by the scenarios agriculture and recreation. The existing situation and the scenario wind turbines can be considered of equal value in terms of landscape preference, recreation preference and ecological value.

Looking at the result from the viewpoint of sustainability, both the production oriented scenarios received greater (scenario agriculture) or equal (scenario wind turbines) scores to the existing situation. The most attractive scenes typically include structural landscape elements such as meadows with orchards, single trees, shrubs as well as grazing farm animals or also more forest.

The least attractive images include scenes where the buildings of the campus ETH Zürich are highly dominant or where there are no structural landscape elements. This should be taken into account when planning future measures in the design and planning of this peri-urban green space.

Acknowledgements

The presented work is part of the GREENSPACE project. It is funded by EU 5th framework and Bundesamt für Bildung und Wissenschaft, Bern. DHM25/Orthophoto © 2004 swisstopo (BA045986).

NEGOTIATING PUBLIC VIEW PROTECTION AND HIGH DENSITY IN URBAN DESIGN

John W. Danahy

Introduction

Over a period of 20 years, the Centre for Landscape Research (CLR) at the University of Toronto developed two key projects. These were based in the National Capital, Ottawa-Gatineau and the City of Toronto. Both projects began as tests to see which elements of urban landscape were possible to visualize and compute with state-of-the-art computer graphics and which elements of landscape can be effectively brought into everyday urban design decision making and planning with visualization (Figure 9.10). The capacities of the machines and the software (e.g. Polytrim) developed by CLR determined the scope of the projects (Danahy and Hoinkes 1995).

9.10 An aerial overview of the University of Toronto Landscape Open Space Master Plan Concept designed by Urban Strategies Inc. Models such as this one become useful if one can freely look about and move around at eye level, comparing visual concerns with systematic economic and planning parameters that affect feasibility and profit (image reproduced in the colour plate section)

Applications

Progressive refinement, comparison and negotiation

CLR's visualization tools have never been used to present a *fait accompli* design. CLR's purpose for exploring visualization was progressive refinement of the tools and process of negotiating urban designs (Danahy and Hoinkes 1995). This meant that a single visualization or animation solution was never the objective and because we were looking at an urban landscape as our unit of study, we never focused on any one development proposal or policy as a finished object. This approach directed the work away from high definition photo-realistic applications of computer graphics visualization. The necessity for linking the work to other data and abstract modelling principles made it more interesting for the people collaborating with CLR to seek multiple changes and time-based viewing of models or phenomena. This observed preference for what we call real-time or dynamic user-directed visualization became a dominant research interest. Our collaborators placed greater value on dynamic visualization than on photo-realistic and animation approaches to visualization (Danahy 2001).

These two urban landscape projects (Ottawa and Toronto) focused on how digital simulation and visualization can help competing interests in an urban landscape find accommodations that minimize the potentially negative effects of both high-density development and visual control mechanisms. Digital media, through iterative built form design and negotiation, can assist an accommodation between 'as-of-right' high densities and desires to minimize visual impact expressed as height and silhouette. The key to influential use of the tools lies in the systematic tie between the visual representations and numeric representations of density and economic value (Figures 9.11 and 9.12).

9.11 Screen snapshots of interactive comparisons between (top) building height; (below) real-time numerical information on economics and planning

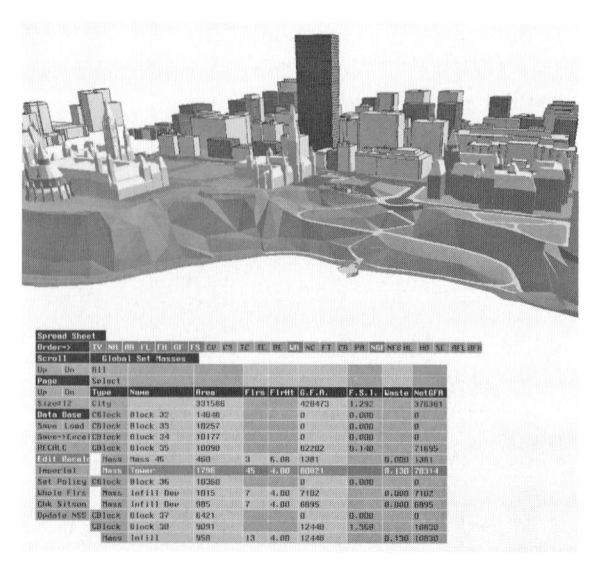

9.12 An illustration of the economic capacity analysis for Ottawa running in real time with interactive modelling and visualizing of impacts on the Parliamentary silhouette

204

The National Capital Case – Ottawa-Gatineau

The first case is the modelling of the Symbolic Core Area of the Canadian National Capital in Ottawa-Gatineau. This work was done in close collaboration with the urban design consulting firm duToit Allsopp Hillier and the National Capital Commission (Allsopp *et al.* 2002). The work began with comparative compositional studies of how to deploy more office space on Parliament Hill without negating the symbolism, meaning and beauty of the Capital landscape. The work progressed to encompass analyses and multiple massing studies aimed at protecting the visual primacy and symbolic integrity of the silhouette of the Parliamentary area and thus retaining the most important experiences of a visitor to the national institutions and landscape of the Symbolic Core Area (Figure 9.13). The symbols consist of a combination of the land promontories extending to the Ottawa River landscape with pavilion buildings set on these bluffs to form relationships with the landscape on all sides. The primary principle regarding development was that it should remain secondary to the reading of the most important symbols. No building should rise above the ridgeline of the Parliament Building and secondary parliamentary buildings such as the East and West Blocks and the Supreme Court should not have background city buildings towering above them. The combined effect of the height of the bluff and the height of symbolic building above the shoreline should be greater than the apparent height of background generic town buildings when read from the controlling viewpoints. The value expressed in this model was that private interest should not be greater than the prominence of the national symbolic core.

The planning challenge was to develop a height control envelope that allowed developers to achieve the maximum allowable density defined in the original zoning bylaw while protecting the Parliamentary Precinct's silhouette. This process entailed creating architectural massing and feasibility studies on the remaining development parcels in the downtown core. The final height control envelopes emerged from an iterative and negotiated process among representatives of the development and citizen organizations (Figures 9.14 and 9.15).

9.13 Visualization was used to systematically show how the tall buildings would change the silhouette and visual prominence of the National Symbols

Applications

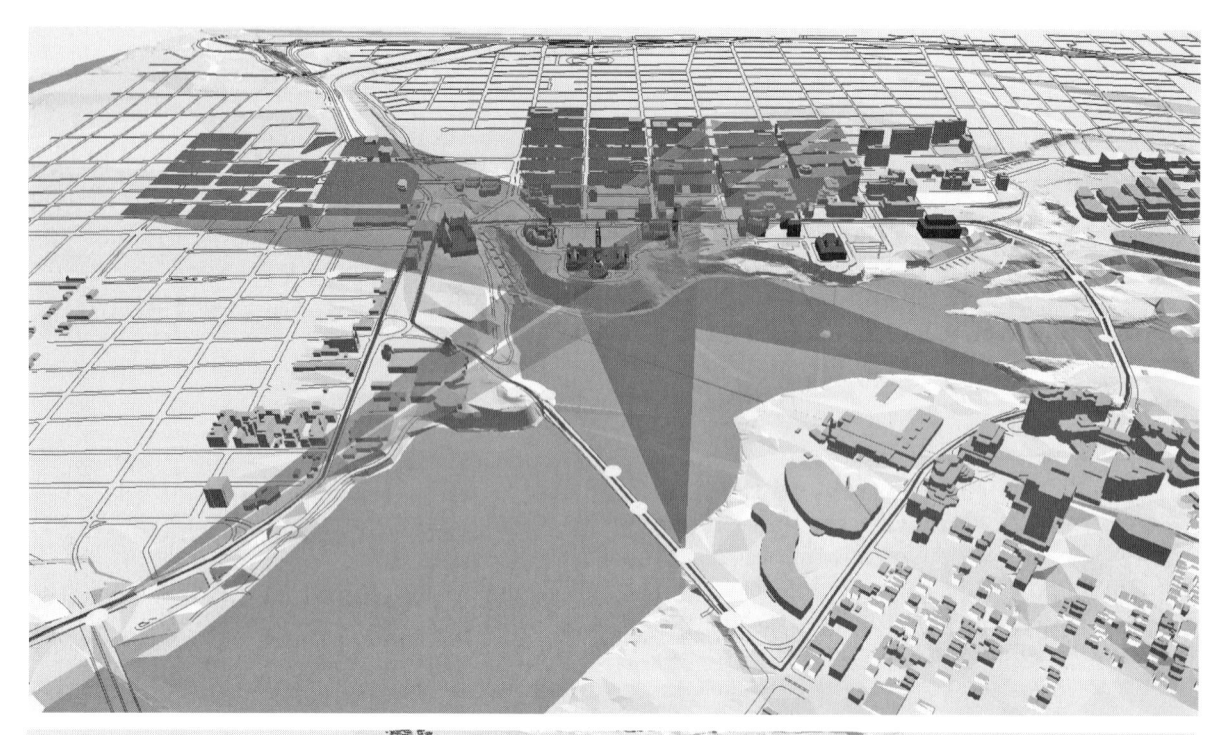

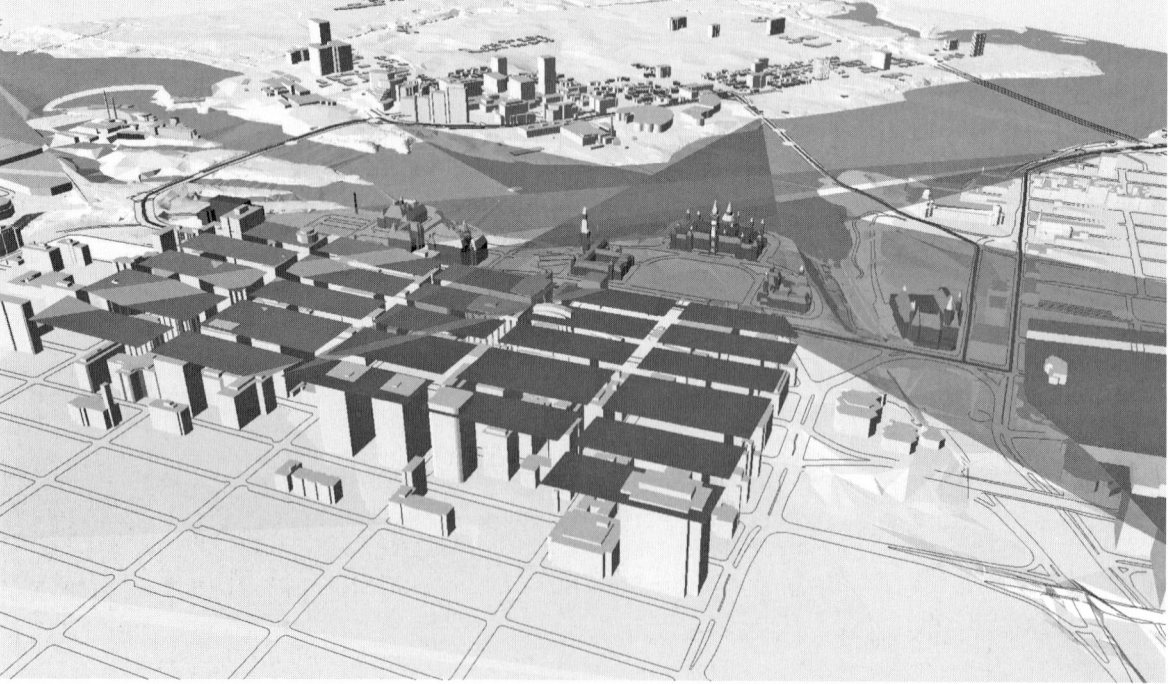

9.14 The controlling views selected as the systematic basis for controlling heights in Ottawa and the key viewpoints (dots) along the ceremonial and interpretive routes (images reproduced in the colour plate section)

Applications in the urban landscape

9.15 The transparent mass in each sequential view indicates the maximum height of future development as seen from the key viewpoints (dots in Figure 9.14) along the ceremonial and interpretive routes

Applications

The Toronto project

The second case study uses two districts of the City of Toronto. One is the western downtown landscape composed of remnants of the Garrison Creek watershed, a heritage battlefield and fort, waterfront, filled harbour lands, rail marshalling yards and corridors and industrial brownfields. These areas are a primary focus for high-density development and expansion of Toronto's downtown core to mitigate the visual dominance of high-rise development at Fort York, Toronto and the Western Waterfront. The second study district is the downtown campus of the University of Toronto.

CLR began imaging possible futures on the Toronto waterfront in 1983. It began with image processing and static computer graphic shaded models. At the time, no one really seemed to want to see the impact that three high-rise residential buildings on the waterfront would exert. Later these buildings became symbolic of a failed planning exercise. Over the years the CLR has tried to intervene with models and imagery propositions that various people have put forward. In the late 1980s, Robert Wright with Stephen Ginsberg worked with the City to explore massing strategies and view corridors to the harbour for the Rail Lands district. In the early 1990s Wright, Hoinkes, Morelli, Ironside and Pedersen worked to develop a landscape model generated from GIS and land record data for a remnant creek system that once formed the western edge of the city. These efforts laid the groundwork for a city model database that Danahy and Lindquist established to illustrate the impacts of proposed high-rise development on the heritage landscape of Fort York and its battlefield. The citizens organization Friends of Fort York asked the CLR to assist them in understanding the nature of the impacts on the experience of users of the Fort imposed by proposed high-rise developments (see Figures 9.16 and 9.17). Next, the panoramic immersive visualization tools were used in an effort to convince the City and developers to not proceed. However, unlike in Ottawa, the mayor and a majority of councilors did not deem the Fort to be a sufficiently significant heritage site to warrant compromising market trends for high-rise condominium towers. Developers' desire for easily marketed units took priority. However, with that macro-value decision made, the City and the developers have been working with the citizens groups and the CLR to search for ways to mitigate the extremes of impact on the visual experience of the Fort. Following from this experiment, the tools are currently being used to assist developers to work with neighbourhood groups early in the development approval process in an effort to find designs that both parties can agree to support.

The key to making this process effective is access to a context model database. With the context model in place, designs can be imported from a consultant's CAD system in a few hours. People are invited to view a walk-through of the surrounding urban landscape of the pro-

Applications in the urban landscape

9.16 Imaging the as yet unbuilt future of the lands surrounding Fort York

9.17 The real-time eye-level panoramic experiences of a person walking about in Fort York were used by citizens to gauge the impact of background and middle ground development for themselves

9.18 A pedestrian-level examination of a foreground view corridor in the first sketch plan produced for Kings College Circle by Andropogon Associates. This scheme took less than two hours to prepare and insert into the campus context model

posal and are free to ask to see and approach the scheme from any position they choose. In addition to the numeric accuracy of the model in terms of building economics, this flexibility gives people an opportunity to 'read' the model for themselves. This characteristic of a real-time system has been central to getting people to participate in negotiations – both in Ottawa and Toronto. The process becomes even more influential if people can make changes to design parameters that are computable, such as moving elements or growing vegetation.

Finally, the most detailed photo texture model we have used in this type of forum is one for the downtown campus district of the University of Toronto. This detailed model has been used to visualize the landscape master plan and specific building interventions and the first phase of landscape architectural designs around the historic core, Kings College Circle (see Figures 9.10 and 9.18).

Conclusion

There is no longer any legitimate excuse not to use real-time visualization to examine the potential effects of density patterns on the visual parameters of a city. The case studies presented here were successful because they link dynamic visual simulation with parameters of development. The sustained experience at the Centre for Landscape Research (CLR) is that visual issues stand a greater chance of influencing the final design of an urban landscape when they can be clearly and systematically presented using digital visualization techniques that are tied to economically important geometric parameters (such as floor area).

The value of the real-time approach to visualization that CLR developed becomes most pronounced when a search for impacts and a search for solutions form the basis of the exercise. If one is simply promoting a given argument, then conventional static and didactically composed cinematic animations will suffice. A real-time approach makes it practical to traverse an extensive virtual landscape. Participants in the process can look at the issues as they wish in order to analyse the problem. Real-time media make it increasingly possible for people to compare their own perceptions with the opinions and decisions presented to them by professionals.

Acknowledgements

The work portrayed in the images here is the product of many people incrementally contributing to the projects over a 20-year period. The list of people involved in these projects is too large to fit in the space available. Remarkably, individuals have come and gone and returned to the effort many times throughout these two decades. The core contributors include staff and volunteers in the CLR, the National Capital Commission, duToit Allsopp Hillier consultants, the City of Ottawa, the City of Toronto, various citizens groups and consulting firms. Links to more information about the people and projects can be found at www.clr.utoronto.ca, www2.clr.utoronto.ca, and the papers referenced here. Support for the basic research agenda has come from various sources including Silicon Graphics Inc., University Research Incentive Fund, The National Capital Commission, Canada Foundation for Innovation, Heritage Canada and Immersion Studios Inc.

COMBINING VISUALIZATION WITH CHOICE EXPERIMENTATION IN THE BUILT ENVIRONMENT

Richard Laing, Anne-Marie Davies and Stephen Scott

The Scott Sutherland School (Robert Gordon University, Aberdeen) undertook a project entitled 'Streetscapes', which aimed to develop methods to assist with public participation in the design of urban streets. The research started with the view that streetscape design is intrinsically complex and textual descriptions of a design might inadequately describe the scene to a non-expert. It was hypothesized that the use of computer-generated images, taken from fully constructed three-dimensional models, would improve this situation and produce more meaningful designs where public participation was taken into account.

Choice experimentation is a non-market valuation technique from the field of environmental economics within which respondents are required to choose between two or more alternatives. Each alternative is statistically derived and contains a mixture of 'attributes'. The results can identify which attributes have influenced choice, and under what circumstances. A typical study would involve asking each respondent to make between six and ten such choices. A choice set typically includes two or more alternatives and a 'do nothing' or 'status quo' option (Blamey *et al.* 1997). It is possible to infer from the results of choice experiments the attributes which significantly influence choice, their implied ranking, the marginal willingness to pay (WTP) for an increase in any significant attributes, and the implied WTP for a design which changes more than one attribute simultaneously (Hanley *et al.* 1998). Within the survey described here, a basic or 'status quo' design for the Castlegate square in Aberdeen was presented, with choices in the experiment itself representing changes to that square.

Choice experiments have a number of benefits which are of particular importance when using visualizations. Adamowicz *et al.* (1995) and Hanley *et al.* (1998) claim that the experimental aspect of choice experiments, means that 'bundles' of attributes can be constructed. When price is included as one of the attributes, it becomes possible to estimate the economic values associated with the other attributes (Boxall *et al.* 1996). It should be noted, however, that choice experiments do not 'require' cost to be included as an attribute, unlike other non-market valuation methods such as contingent valuation. Finally, Adamowicz *et al.* (1995), Morrison *et al.* (1996) and Hanley *et al.* (1998) argue that the issue of embedding, a commonly discussed problem in contingent valuation studies, is overcome through choice experiments. That is, the 'true' value of a good or attribute might be masked due to the context within which it has been displayed. Kahneman and Knetch (1992) provide a discussion of this form of bias. As choice experiments are formed using varying levels of attributes, different subsets of goods (including substitutes) are essentially incorporated within their design.

Although environmental economics is regarded as an established field within economics, the quantification of socio-economic impact is rarely attempted, in the UK at least, and very rarely in relation to architectural design (Chadwick 2002). The intention of the Streetscapes project was to signal a method through which environmental economics might aid the design development process. Therefore, throughout the study, the high quality of images was not in any way compromised by a lack of rigour or validity in the experimental approach.

Visualization

As stated, this project investigated the extent to which visualizations could be used effectively within a choice experiment. Although many aspects of a space are non-visual to some extent (e.g. temperature, noise), a number of prominent recent urban design guides focus very much on making physical changes to space. For this particular study, the major attributes which were ultimately investigated all concerned the manipulation and configuration of physical objects within a space. To that end, the use of AutoCAD and 3D Studio Max was appropriate for the modelling and rendering. For subsequent projects, the team has successfully used these packages to model and visualize more 'environmental' effects, although the packages are more limited, perhaps, when considering ecological details such as plant growth.

Focus groups

A case study area within Aberdeen was selected for study, and a series of focus groups were completed, including meetings with local residents, design students and built-environment professionals. The main purpose of these meetings was to identify possible uses for the case study area, and to suggest attributes which could improve the square.

A major finding from the focus groups was that participants generally responded more favourably towards design options for the area which reflected a change of use, rather than simply considering individual streetscaping objects. For example, respondents rarely suggested planting new trees in the absence of a strong rationale. Rather, an emphasis was placed on the social environment, or the impact of physical changes on peoples' behaviour within the square.

From these focus groups, a number of possible arrangements for the space were designed, using various combinations of railings or bol-

Applications

lards, trees and planters, benches and bins, paving, and lighting. Following further modelling and rendering work in 3D Studio Max, a series of high quality visual images were produced for use in the choice experiment. For each image, clear details were recorded about the environment being shown, and the configuration for each of the variable objects. Sample images from the model are shown in Figure 9.19, with images taken from the study itself shown in Figure 9.20.

9.19 Sample images from the model: (top pair) object details from the model; (middle pair) images illustrating the modelling process; and (bottom) image from the final model

Applications in the urban landscape

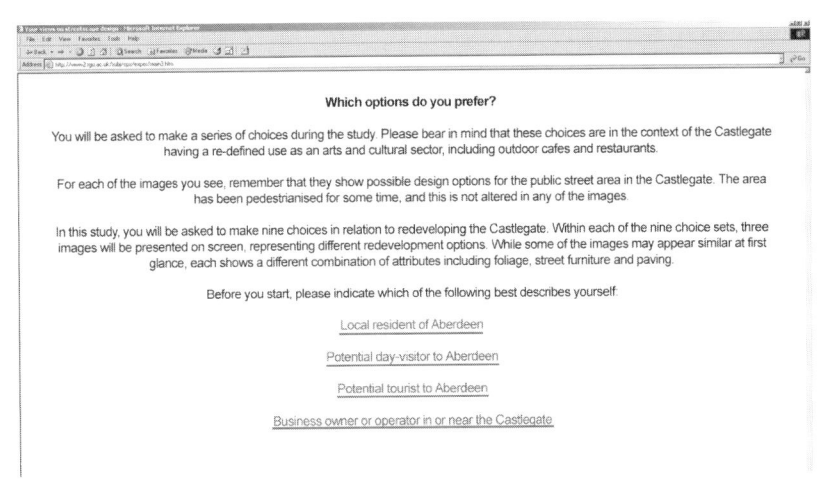

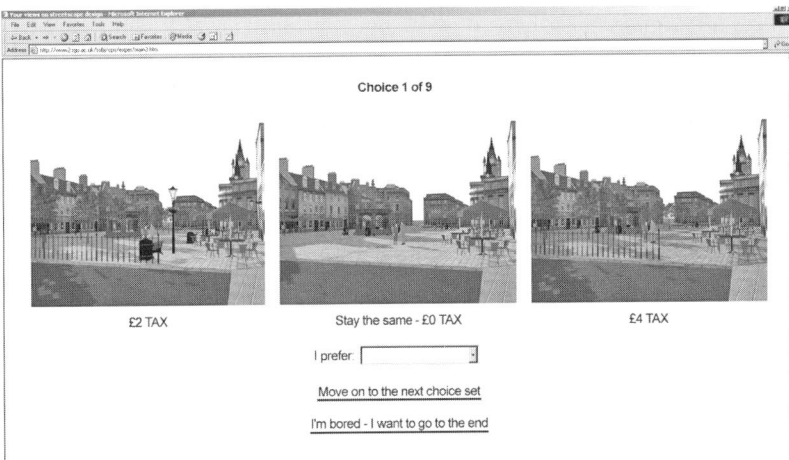

9.20 Sample images from the study: (top and middle) sample pages from the main study: (bottom) clickable enlarged image from the study

Experimental work

The choice experiment was completed in two stages, where the main differences were concerned with delivery of the material and control over the technology. The first phase of the study was implemented through an entirely Internet-based experiment. Following that study, the choice experiment was replicated on a stand-alone machine, thus providing further data to the study as a whole, and also allowing for a comparison between the two forms of data transfer. In addition to the choice experiment itself, several open-ended questions were included to give respondents the opportunity to make any comments regarding the study, as well as provide feedback. Questions were asked about image quality, payment levels, ease of answering the questions, and the overall quality of the study.

In response to the Internet-based study, nearly 20 per cent of respondents commented on image size, brightness, clarity and download time. Despite the survey including provision for the standardization of monitor settings, there still appeared to be some variation between users. While this problem may have been the result of respondents not adjusting their monitors at the beginning of the study in accordance with instructions, it is clearly a problem for any Internet-based study where visual images are paramount to the questions.

In order to minimize download times, the Internet study used smaller 'near-thumbnail' sized images, which could be opened to reveal a higher definition image. Faster Internet connections would ease the need for such measures, and facilitate a more rapid transfer of data in both directions.

The study also contained a 'cost' variable which raised interest from respondents, as a tax levy was attached to each design alternative. This is common to non-market valuation studies, but often carries ramifications with regard to appropriateness, cost level, and even ethical considerations. Of those respondents who commented on the cost levels, decisions were split between those who felt the suggested costs were too high and those who felt the levels were reasonable. Some also commented that money raised through taxation might be better spent on alternative projects, including traffic management within Aberdeen.

The main differences between the Internet and the standalone studies were the sizes of the images presented to respondents and the amount of information provided. The standalone study was conducted on one standalone computer, which made it easier to control some of the issues discussed above. For example, issues concerning individual monitors (such as screen size and brightness) were eliminated.

A number of the Internet study respondents argued that more information regarding the case study space and the design alternatives was required, claiming that it was difficult for them to identify some of the changes proposed from images alone. This was addressed specifically in the standalone stage by providing text descriptions of the changes made in addition to the images. In the standalone study the percentage of respondents stating that images were 'too similar' or of 'poor quality' was reduced by almost 70 per cent.

From a methodological perspective, the most frequent comments from the standalone study tended to focus on the attributes displayed within the images, including the configuration of trees, paving type, and so on. Differences in the comments made by the Internet and standalone respondents suggest that when larger, brighter images are provided (which are fast to download), respondents are better able to concentrate on image content rather than on technical issues. This would suggest that using a standalone computer is more suitable for presenting detailed images to respondents, and some level of text to highlight differences in the images is required. This situation may change, of course, as standard Internet connections become faster.

It is well accepted that as the number of possible choice options increases, so do the cognitive demands associated with making choices, and thus the level of confusion increases (Blamey *et al.* 1997). Analysed results from the study suggest that respondents did identify modelled attributes, and that changes in the model did indeed influence choices.

Image realism and related issues

In addition to the open-ended questions that allowed respondents to make comments about the questionnaire, several closed-ended questions were also asked. These included:

- How realistic do the images look?
- How easy or difficult was the survey to answer?

A quarter of the 172 Internet respondents said that the images looked 'very realistic', compared with nearly half (43 per cent) of the 65 standalone respondents. Similarly, less than 2 per cent of the standalone respondents thought that the images looked 'very unrealistic', compared with 5 per cent of the Internet respondents. Almost 60 per cent of the standalone respondents and 46 per cent of the Internet respondents found the survey 'very easy' to answer.

Overall, these results suggest that the majority of respondents thought that the images were realistic (Internet: 81 per cent, standalone: 88 per cent) and that the questionnaire was easy to answer (Internet: 77 per cent, standalone: 88 per cent).

Although this study used only still images taken from the model, the research team was well aware that there is potential for the use of far more interactive methodologies. Advances in Internet technology in the past 2–3 years, particularly with reference to the use of Flash, VRML and technologies developed for online gaming, suggest that there may be a desire within the research community to use such techniques in research. It is essential that care be exercised when using visual images as part of such studies, particularly where questions relate to perception, as control over image quality and delivery will inevitably vary greatly between computers.

Attribute-related results

An aim of this study was to investigate how physical 'attributes' could be visualized within choice experiments. It was essential, therefore, that the study demonstrated that those people taking part were able to distinguish between images, and that the 'attributes' shown have somehow influenced the choices made.

The physical changes shown in the study images related to railings, trees, benches, paving, lighting and plants. In addition, all choices were allocated a 'price', which would be paid through taxes.

The study results showed that trees had the biggest effect on choice. A significant addition of trees to the square was valued most highly by respondents, with current levels indicated as poor (negative) in the responses received. Respondents also preferred those scenes where bollards had replaced the existing railings.

Although benches and bins appeared to have a negative effect on value in their current layout, changes were insignificant and did not appear to alter choices. A change of paving colour from grey to pink granite was, however, seen to have a positive influence. Additional lighting in the square was judged to have a positive effect, (although the study did not test the effect during night-time hours). The addition of planters alleviated a poor valuation of the current layout, although the effect was not significant in comparison to that caused by trees. Finally, cost did appear to have a significant effect on choice, in the sense that respondents would prefer to pay lower rates.

As most of the attributes modelled and visualized were significant to choices, the future use of this method in similar studies is justified. Respondents were clearly able to distinguish between images, and understand scenarios when these were presented to them in an aggregated and highly visual format. Although care needs to be taken to match the experimental study area to the method, the approach described has great potential for future studies in both the built and natural environments.

Conclusion

This project was concerned with an appraisal of the validity of computer visualization within studies concerning the perception of urban space. Although a limited number of choice experiment studies have in the past attempted to use images as part of the survey, this research used images to convey almost all aspects of the choice scenarios. The ability to use computer-generated (as opposed to manipulated) images of a near photo-realistic quality also reflects advances in technology which should be investigated in an objective manner. By using fully computer-generated environments, the researchers were able to exercise far greater control over the contents of scenes, than if using photographs.

The main findings of this study can be categorized under two headings, namely, the effectiveness of the 'visualization' within choice experiments, and the technical limitations. With regard to the choice experiment, results from the research (Davies and Laing 2002a; 2002b) indicated that respondents were able to drive the survey design (through participation in focus groups) and also understand the content of the images, and that a number of the modelled objects were found to have a significant influence on choice. With regard to technical limitations, the research used an Internet browser to collect data for reasons of widespread compatibility and inclusiveness. However, due to current limitations and variation between user equipment and connection speeds, some respondents suggested that they were distracted from the survey by technical barriers. The subsequent standalone study served to show that most of these barriers can be overcome, and that the method appears to offer great potential for the future.

The project results suggest that the use of 'virtual' environments holds potential for greater use within non-market valuation. It is anticipated that research using immersive and interactive technologies could be used as a tool to highlight or identify those key aspects of an environment that critically influence choice and preference. As the level of 'realism' possible in virtual models increases, this must be balanced against the central aims of a research project, and the need for methodological rigour. Davies and Laing (2003) explore the subject of image contents and the influence on overall choice in greater depth.

The intention of this work was to encourage creativity, stimulate discussion, refine ideas and ultimately produce designs which optimize the outcomes for all parties involved. The successful use of focus groups to define the problem area, followed by the innovative use of visualization within the data collection represents an interesting contribution to several fields, and should be developed further.

Acknowledgements

The work reported in this paper was jointly funded by Scottish Enterprise and The Robert Gordon University. Further details of the study can be found at www.rgu.ac.uk/sss

Part 3

Prospects

CHAPTER 10

VISUALIZATION PROSPECTS

INTRODUCTION

Ian D. Bishop and Eckart Lange

The applications described in the preceding chapters are all based on current technology or current procedures made more efficient. In this chapter we try to look forward to emerging technologies and also consider more closely the way in which these technologies may be integrated with emerging principles for public participation in environmental planning.

Considering first technology: several of our authors have stressed the importance of seeing environmental changes in context. That is, seeing the whole environment including those elements that are new or changing.

This raises the question: why should we take the trouble to model those components of the landscape that are not changing? Most of a forest, for example, may remain standing under some scenario of change. Nevertheless with the techniques used in most of our case studies we are required to construct a detailed model of this forest. Why should we not use imagery of the existing landscape and only use computer graphics to represent the areas of change? This is the concept behind augmented reality and its application to landscapes and environmental modelling. This was also the concept behind photomontage techniques used in the 1960s and 1970s with computer graphics superimposed on photographs (Bureau of Land Management, 1980). Indeed, we can treat photomontage as one special case of augmented reality. However, the use of still images does not permit objects in the view to move (e.g. cars, trees in the wind) nor the camera to move. The ultimate augmented reality (and the only type according to some authors – e.g. Azuma *et al.* 2001) is for the graphic superposition to occur in real time via a partial head-mounted display. This allows change to be seen directly in conjunction with existing conditions from any location, and in any direction, to which the user moves.

This sounds excellent in principle but there are technical and procedural difficulties that must be overcome first. The following sections of this chapter show that the process of dealing with these issues is well under way.

The first contribution deals with video as the source of existing images. This is part way from the simple case of photomontage to the complexities of real-time AR. Eihachiro Nakamae and his colleagues (see 'Compositing of Computer Graphics with Landscape Video Sequences', p. 226) consider that a video sequence can provide an excellent representation of existing conditions. The sequence might involve the camera panning across the landscape, zooming on an object of interest or being driven along a street. They describe the means through which they have overcome a range of difficulties related to camera location, occlusion, lighting and moving objects.

Piekarski and Thomas (see 'Future use of augmented reality for environmental and landscape planners', p. 234) then take us into the expanding world of real-time AR and look specifically at its potential roles in landscape and environmental planning: for example, building design while on site.

In relation to public participation nearly all of the preceding contributions in the applications chapters have stressed the potential of the technology in assessment of public attitudes or in organized workshop situations. However, in none of the earlier cases was the focus on a web-based information system for public consumption. While Andrew Lovett (see Chapter 7, 'Designing, visualizing and evaluating sustainable agricultural landscapes', p. 136) used VRML for his models of the Thames valley, the delivery was primarily by CD as farmers were interviewed on an individual basis. On the other hand, Nathan Perkins and Steve Barnhart (see 'Visualization and participatory decision making', p. 241) describes two projects based entirely on web-delivery. This approach will certainly be increasingly used in the future as advances in broadband access make viewing of more complex models possible.

Most applications combining GIS with visualization describe a comparatively simple process of generating three-dimensional models from two-dimensional data. This may involve extrusion of buildings, incorporation of road alignments into a DTM, population of areas with trees or transmission towers along an easement. Verbree *et al.* (1999) largely automated this process in an urban context. Perrin *et al.* (2001) describe their IMAGIS approach to landscape modelling from geographic information.

The greater challenge is to make this linkage interactive such that any manipulation of data in the GIS is immediately reflected in the three-dimensional model. Jepson *et al.* (2001) review the extensive work which produced Virtual Los Angeles, and which also led the way in forming a two-way network link between a GIS (ArcView) and a real-time visualization system (OpenGL Performer). Van Maran and Germs (1999) have linked the WorldToolKit (WTK) from Sense8 with ESRI's Spatial Data Engine (ArcSDE) to allow for interactive enquiry and editing of the spatial data from within the three-dimensional view. Stock and Bishop (see 'Helping rural communities envision their future', p. 145) described their process for integration of GIS with a Performer based visualization system and handheld computers for community input in a workshop setting.

Integrated systems which offer both GIS and visualization functionality increase the scope for collaborative work between planning and management professionals and better communication between professionals and the public. In the work described by Michael Kwartler (see 'Visualization in support of public participation', p. 251) the emphasis is on software development specifically for community-based planning. CommunityViz is an example of visualization development being closely coupled to GIS and is based on the product Creator from MultiGen.

This chapter concludes with an attempt to overview these trends and also to identify the major challenges facing visualization and visualizers in the immediate future.

COMPOSITING OF COMPUTER GRAPHICS WITH LANDSCAPE VIDEO SEQUENCES

Eihachiro Nakamae, Xueying Qin and Katsumi Tadamura

Introduction

In this section rendering techniques of outdoor scenes, using computer-generated still images composited with video-sequence frames, are discussed. The techniques, which are intended for use in visual environmental assessment, can be classified into two categories (Nakamae 2001; Nakamae *et al.* 2001). The first uses a panoramic landscape image created from panned video-sequence frames and the second a landscape video sequence. Both are composited with computer-generated still images. Geometrical and optical consistency are the main problems in making the panoramas or video-sequences match well with the computer images.

Panoramic images

For environmental landscape assessments, panoramic observation with a wide screen is much more useful than a set of standard size still images. By observing the panoramic landscape image composited with the computer-generated still images, observers can evaluate the design and harmony of the planned structures in detail. The process is easy and inexpensive and yields a wide landscape view.

Problems for compositing a panoramic image from video sequence frames

A number of techniques have been developed for capturing pure panoramic images of a real-world scene. The image mosaic method uses several regular photos or video frames (Szeliski 1996; Shum and Szeliski 1998). These images are aligned and composed by using an image mosaic or 'stitching' algorithm, in this case optical effects are not too severe, i.e. the effects of auto-iris, vignetting and interlacing are ignored and the geometrical mismatch due to camera lens and/or charge-coupled device (CCD) distortion is dealt with by employing the de-ghosting method or local alignment. These problems, however, cannot be ignored for compositing a panoramic image from continuous video sequence images as discussed in the following sections.

Making a panoramic background image

A panoramic image can be created by the following processes (Nakamae *et al*. 1998; Qin *et al*. 1999). Set an appropriate panoramic image space accounting for the range of the video-sequence frames. Extract the camera parameters of each video-sequence frame (see below) selected at several frame intervals to avoid the influence of vignetting (image intensity drop at edges). Rectify the distortion of the selected frame images, which originates in the video camera lens/CCD, until the maximum error is within one pixel or less. Remove every moving object existing in each frame because the moving objects in a video sequence might yield a kind of motion blur in the panoramic still image, e.g. when a moving car's motion synchronizes with the panning speed, the car must be drawn like a straight bar (refer to Qin *et al*. 1999 for details). Modify the interlaced images produced by an interlaced digital-video camera in order to prevent jags from appearing in each frame. Select one appropriate exposure frame and adjust the luminance of the other frame images to it, because when an auto iris video camera is panned, the exposure of each frame changes depending on its incident light. Cut out the midrange of each frame with the width corresponding to the rotating angle of the video camera, and put it in the appropriate place in the panoramic image space. After completing these pre-processes, interactively paste the moving objects onto the respective frames.

In order to harmonize the computer-generated images of structures with a panoramic background image in daylight, the rendering algorithm should account for the following optical phenomena: the incident lights, including direct sunlight and skylight, and the phenomena of light scattering and absorption caused by air molecules and aerosols such as fog and clouds (Nishita *et al*. 1996); the shadows cast by not only solid obstacles but also clouds (Tadamura *et al*. 1999); and the spectral characteristics of reflection, refraction and transparency of water. Modelling realistic natural objects such as trees (Qin *et al*. 2003) is also indispensable; the illumination circumstance and projection coordinate system of the computer-generated images should be matched with those of the background panoramic image (Nakamae *et al*. 1986) and the computer-generated images may be partially hidden behind the foreground image elements. A panoramic scene developed in this way is shown in Figure 10.1. Both composited panoramic images, Figures 10.1(c) and (d), are matched well in both geometry and optics.

Prospects

a

b

c

d

10.1 Panoramic scenes: (a) original panoramic scene composited from a panning video sequence taken in the evening; (b) panoramic scene of computer-generated images corresponding to (a); (c) composited panoramic image of (a) and (b); (d) panoramic image composited from a panning video sequence taken in the afternoon (images reproduced in the colour plate section)

Composited video sequence with computer-generated still images

Compositing video sequence taken on a tripod or one's shoulder and computer-generated still images

Although the panoramic landscape images displayed on a wide screen are useful, objects which are normally in motion, such as cars, people and waves, are frozen. This approach also has difficulties in providing sufficient resolution for zooming in.

By employing a fully automatic camera-tracking algorithm, a panned, tilted, rolled and/or zoomed video-sequence composited with computer-generated still images can provide excellent vivid presence at low cost

(Qin *et al.* 2002). It is especially attractive for making amusement movies, while for visual environmental assessment relatively slow panning and/or zooming are recommended to accommodate an accurate, objective portrayal.

If the video-sequence frames and the computer-generated images were composited directly, the images might move and shake in the video sequence due to the distortion of the optical lens/CCD and interlacing of video-sequence images. These problems become more serious when the views change slowly as is usual for landscape environmental assessment. Observers can easily notice the jittering of computer-generated still images, even if the jittering is as little as one pixel. To solve these problems the following processes are recommended.

Basic geometrical relationship

The coordinate systems of computer-generated objects, camera and video-sequence should be accurately matched by employing the following:

1. any three-dimensional point vector in the world coordinate system is projected onto the perspective projection plane of the video camera;
2. the lens/CCD distortion of the video camera is expressed as a function of the focal length; and
3. any point vector of a video-frame image is a two-dimensional scale function, which expresses the relation of the successive frames, given by the functions mentioned above and a rotating matrix with tri-axes.

Computer-generated objects are projected onto their own projection plane. In order to map their images onto the video-sequence frames, the transform matrix to match the computer-generated object coordinate system to the coordinate system of the first frame of the video sequence should be prepared by using a terrain map. Rendered images should be modified to match each video-sequence frame by using the camera parameters and taking into account interlacing, distortion (Li and Lavest 1996) and perspective projection.

Automatic camera tracking

Camera parameters are recovered by analysing each frame of a video sequence. In this process their accuracy is strongly influenced by the moving objects in the scenes. However, the parameters can be automatically recovered with high accuracy (without tracking the moving objects) by the following processes.

Every frame is divided into sub-images and, in addition, each sub-image is divided into cells as shown in Figure 10.2, and the sub-images with rich textures are selected for further use. For robust camera

Prospects

10.2 Selected and unselected subimages. The sub-images and cells surrounded by dark lines have a poor-texture; those surrounded by white boxes are rich-texture subimages

parameter construction, the following two-step process removes sub-images containing moving objects automatically.

- Step 1 – Search for panning and tilting parameters only, by tracking every rich textured sub-image individually, because two parameters processing is faster and more robust. The difference of the lens distortions in the sub-images can be ignored because the distortion between the adjacent frames is similar. Calculate the camera movement from the translation of each sub-image centre. Remove the sub-images that seem to have moved furthest because they must be moving objects or noise. Then obtain the convergent parameters using least squares.
- Step 2 – Search the parameters for rolling and zooming by using the eligible sub-images in step 1. Detect the positions in every frame related to the sub-image centre and its corresponding point in the consecutive frame. Again, obtain the convergent parameters using least squares. Examine if the deviation exceeds a certain tolerance, if necessary set these parameters as the new initial values of step 1 and repeat the process. Detect the brightness change between these two frames by employing the remaining sub-images (Nakamae *et al.* 1998; Debevec *et al.* 1998).

Despite the cautious detection of parameters using the two-step process, the accuracy is still not sufficient to keep the tracking errors small enough over the entire set of video-sequence frames because of the influence of moving objects and noise in the frames, the lack of consideration of lens distortion, and changes of each successive frame's focal plane. Even if the error in each frame is only 0.2 pixels, after one second, the accumulated error becomes six pixels. If the accumulation of displacement of parameters exceeds a certain tolerance then detect the camera parameters again by employing the two-step process described above. With the increased iterations and shorter intervals, a more precise result is obtained.

For a panning and tilting video sequence, it is necessary to detect only two parameters; they are usually stable and can be set with a large refinement interval. However, for a tilting, rolling and/or zooming video sequence, more frequent iteration is needed.

Compositing images with video-sequence frames

After calculating all the camera parameters and matching the computer-generated images into the first video-sequence frame image, the images can be automatically composited with all the remaining frame images by considering the following conditions in terms of both geometry and optics.

Using the method described above, the computer-generated still images can be modified to cope with the distortion. Then they are interlaced in every frame. After these processes, they are composited with every video-sequence frame image by taking account of anti-aliasing and calibrating brightness to match with each video-sequence frame.

In order to show the necessity of precise refinement, Figure 10.3 shows an example of intentional simultaneous panning, tilting, rolling and zooming filmed from a person's shoulder with a 15-second video sequence. The computer-generated images, the towers, the bridge and the roadside trees, are well fixed in Figure 10.3(e). Even though it is somewhat difficult to recognize the moving car, it gives excellent vivid presence in the real video sequence.

Compositing video sequence filmed with a moving camera and computer-generated images

Compared with the techniques in the previous sections, compositing a video sequence filmed with a moving camera and computer-generated images can give greater visual immersion. Recent advances in pattern recognition algorithms for calculating moving camera parameters have replaced the need for a costly camera controller and thus become popular in the entertainment industry. However, in landscape environmental assessment for the images to be as precise as those of the previous examples, it is necessary to re-render the scenes from a different position in each frame. This also requires calculating the shadows each frame, which is extremely costly because much more precise modelling of numerous background objects is necessary.

Prospects

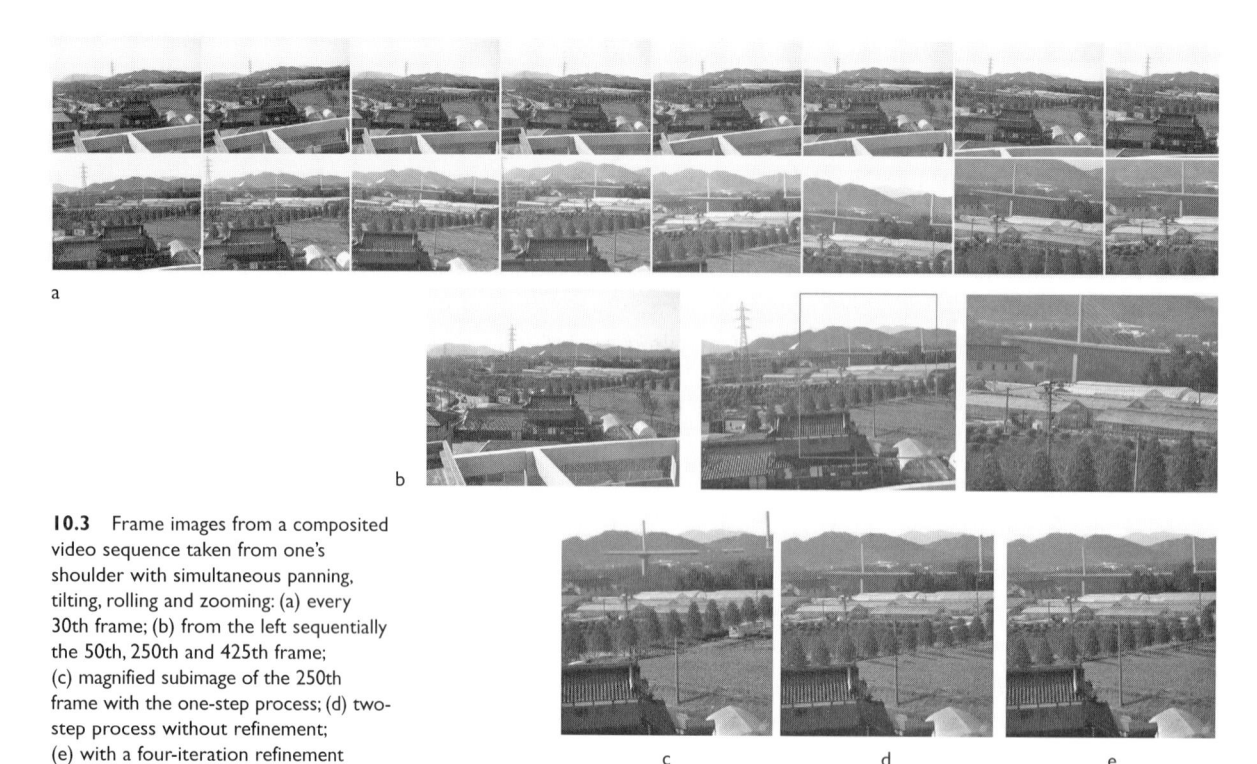

10.3 Frame images from a composited video sequence taken from one's shoulder with simultaneous panning, tilting, rolling and zooming: (a) every 30th frame; (b) from the left sequentially the 50th, 250th and 425th frame; (c) magnified subimage of the 250th frame with the one-step process; (d) two-step process without refinement; (e) with a four-iteration refinement

Other techniques and future problems

Nowadays, the environment in which designers can use computer graphics techniques directly or participate as a postprocessor, is rapidly blossoming. Handy digital high-definition video cameras (e.g. 210 million-pixel CCD) have been released, and many display systems of single/stereo images with/without glasses (e.g. from OmniMax, CAVEs, cinemascope and three screen displays of various sizes to portable spherical screens) are on the market. However, in the achievement of complete immersion, the following problems still exist.

Current CAVEs and similar instruments used for walk-through are unable to effectively generate true landscape images for conveying accurate scenes in terms of both geometry and optics. Visual cues to distance and size and the needed brightness discrimination of displays are lacking. In the former, the effective distance of binocular parallax and convergence is, at most, tens of metres, and in the latter the maximum number of brightness levels is usually only 256 compared with a daylight range of tens of thousands.

The advancement of CAVEs, which enable precise real-time photorealistic rendering, may give observers more visual immersion by solving the following problems. For cues to distance and scale, CAVEs with high-definition stereo displays have the possibility of adding the effects of motion parallax and improved focus adjustment of the observer's crystalline lens. For the optical problems, developing fast rendering with high quality is an urgent problem, because if the problem of computing time is excluded, the rendering techniques for projecting, shading, shadowing, texturing and calculating contrast of brightness depending on the weather are already mature. At present, there is no apparent solution for the absolute brightness problem, even though a projector with 622 million pixels and a light source the same as for commercial cine-projectors is on the market for domestic use.

Viewed from perceptual psychology, the development approach should be quite interdisciplinary, i.e. including computer graphics, computer vision and computer science as well as robotics and acoustics.

FUTURE USE OF AUGMENTED REALITY FOR ENVIRONMENTAL AND LANDSCAPE PLANNERS

Wayne Piekarski and Bruce H. Thomas

Introduction

Key to environmental and landscape planning is the visualization of changes to existing features. Current technologies for visualization of such changes include hand-drawn artist renditions, three-dimensional models viewed on standard desktop computers, and three-dimensional models viewed on immersive virtual reality systems. These three technologies all lack the ability to provide the planner with a first person perspective of the changes in situ. In this section we propose the use of augmented reality (AR) as a means of providing the planner with such ability. The AR overlay contains virtual objects such as buildings and trees that appear to exist in the physical world (Azuma, R. 1997; Azuma *et al.* 2001; Azuma, R.T. 1997).

We believe the ability to view design options on site for environmental and landscape planners is a powerful tool. This tool would help the planners to visualize options overlaid upon the physical world, making the impact of the different designs more tangible to the planner. The interplay between each of the options and the existing environment and landscape is much more obvious to the planner when viewed using AR. Some options are difficult to visualize without all the surrounding landscape and environmental features, such as colour or space. The ability to move freely in the environment allows the user to understand sizes and distances in a similar way to perceptions in the physical world. Changing the user's viewpoint is as easy as walking to a new location. The visualization of such models on a traditional workstation using a mouse to change viewpoints does little to help the user understand such concepts. As a second example, imagine that a planner is reviewing the installation of a new walkway, and the walkway intersects a drainage ditch, which is not depicted on the surveyed maps or site drawings. It would be easy for the planner to detect this problem with an AR tool, reducing both costs and time.

This section will first provide a brief introduction to the concept of AR and the required technologies to support AR in an outdoor environment. The section goes on to present two examples of visualization: landscape design and building design. Finally, some concluding remarks are offered on possible future directions of the technology.

Visualization prospects

Technology

Augmented reality is the registration of projected computer-generated images over a user's view of the physical world. With this extra information presented to the user, the physical world can be enhanced or augmented beyond the user's normal experience. The addition of information that is spatially located relative to the user can help to improve their understanding of size, shape, location and colours of objects such as trees, shrubs, walkways and walls. The schematic diagram in Figure 10.4 depicts how a see-through HMD (used to produce AR images for the user) can be conceptualized.

To provide AR images while the user is mobile, there are a number of technologies that must be integrated. Working first from what the user sees, the AR images are presented through a head-mounted display, as shown in front of the user's eyes in Figure 10.5. We use a Sony Glasstron head-mounted display that is mounted on a helmet with a

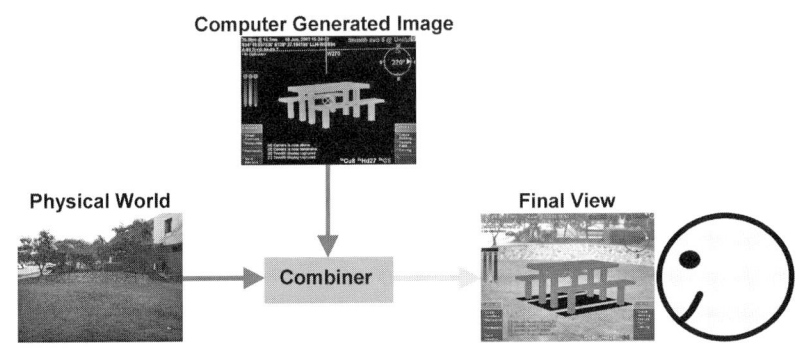

10.4 Overview of augmented reality

10.5 Tinmith-Endeavour outdoor augmented reality mobile computing system (images reproduced in the colour plate section)

FireFly 1394 camera for live video input. To generate virtual images that align with the physical world, the computing system must know the user's position and orientation of view. The technologies we employ are an Intersense IS-300 hybrid magnetic and gyroscopic tracker for orientation sensing, and a Trimble Ag132 GPS with an accuracy of 50 cm for position sensing. This sensor data is processed to render the final view for the head-mounted display. We use a Pentium-III 1.2 GHz laptop with an Nvidia GeForce2 graphics chipset to operate our applications. A single user must be able to carry all this necessary equipment; so a custom backpack computer was designed to support our research, named Tinmith-Endeavour, and is shown in Figure 10.5. The backpack weighs approximately 15 kg and operates for 2 hours. We implemented Tinmith-Endeavour with as many off-the-shelf components as possible, but in the future this could be made smaller and lighter with new technologies and custom-built components.

The techniques discussed in this section have all been implemented and tested in an outdoor environment, unless noted otherwise as future work. We have developed a complete and useable outdoor augmented reality application known as Tinmith-Metro, which allows users to visualize and capture three-dimensional outdoor geometry in the field (Piekarski and Thomas 2001). To implement this application, we use our flexible Tinmith-evo5 software architecture (Piekarski and Thomas 2003), which is a complete toolkit for the development of high performance three-dimensional virtual environment applications.

Landscape design

There are two basic methods of modelling that we have been investigating. The first involves the placement of predefined graphical objects and then modifying their position, scale and rotation transformations. The second is the specification of planar shapes through the placement of vertices using the body.

Placement of predefined graphical objects

The placement of prefabricated objects is performed at the feet of the user as they stand in the environment, such as the table, person and tree shown in Figure 10.6. This method works well when objects that are to be created are known in advance or an approximate graphical model may be used. The physical movement of the user is used to control the object placement. Instantiating objects at the user's feet makes tracking of other body parts such as the hands unnecessary to simplify the equipment required.

Once a predefined graphical object has been placed in the environment, the user can change the object's size, position and orientation.

Visualization prospects

10.6 Example of a predefined table, person, and tree placed in the environment

Based on some initial inspirations from Bolt (1980) and guidelines by Brooks (1988), the user interface makes use of tracked gloves so the user can point at objects and interact with them. Users may perform selection operations using a single hand with a cursor that is projected onto the three-dimensional environment.

At the start of the operation, the user selects an object. As the user's hand moves, this performs translation and maintains the same object to user distance. If the user moves their body during the translation operation then the object will be moved in the same direction as this motion. If the user changes their head direction then the object will rotate around with the user's head.

Scale and rotate operations may be performed more naturally through the use of two-handed input, which was first pioneered by researchers such as Buxton and Myers (1986) and Hinckley (1994). Scaling operations are initiated by selecting an object with the non-dominant hand, and the dominant hand controls the stretching of the object. Rotation operations are performed similarly by selecting an object with the non-dominant hand, with the angle being controlled by the relative position of the dominant hand. Figure 10.7 shows an animation with a sequence of frames of a tree being rotated about the head cursor using two-handed input.

Prospects

10.7 Rotation operation being performed using two-handed interactions

The user interface has a menuing system designed around an arrangement of finger presses with a special set of gloves. The fingers on the user's gloves map to a set of menu options, and the user can navigate through a hierarchy by pressing their fingers against the thumbs. While each of the previously described transformations use interactive pointing by the user, fixed increments controlled by the menu are also possible. The nudge series of commands can move objects in one metre or other fixed translations, rotate in fixed angles along the axes, and scale in fixed increments. The scale and rotate nudge operations also provide the ability to scale objects away from the user or rotate around the ground plane normal, which is otherwise not possible using image plane based inputs.

Specification of planar shapes

To mark out a region on the ground's surface, the user walks along the perimeter of that region they wish to model and places markers at points of interest under the feet. This is also similar to creating waypoints when using a GPS system while walking. When the user has walked completely around the area they wish to specify, a closed perimeter is formed and converted into a closed graphical region. While the initial perimeter defined is a thin polygon, it can be infinitely extruded up to define a solid building outline, or extruded down to approximate the bottom of a lake or river. This technique has been used to model roads, parking lots, grassy areas and other concave outline style shapes. Paths and navigation routes may also be defined, but treated as a line segment instead of a polygon.

Visualization prospects

Building design

The ability to define large building structures is also useful in helping planners in the visualization process. This example highlights the modelling techniques with the construction of an abstract building in an outdoor environment. The user walks outside to an empty piece of land and creates a landscape that they would like to preview and perhaps construct in the future. As an added feature, this model may be viewed on an indoor workstation either in real time during construction or at a later time. Applications for creative purposes such as an abstract art or landscape gardening design tools may be brought into operation during this design phase.

Figure 10.8 shows the user's augmented reality view of this example through a see-though head mounted display at the end of the construction process, and Figure 10.9 shows the same example as an immersive virtual reality view. The first step in creating this new building is to create the perimeter of the building shape by walking around the building site and placing down markers at key positions of the building, forming a flat outline. Next, the outline is extruded upwards into a solid three-dimensional shape. The user stipulates the shape of the roof by specifying various control points. After the overall roof structure is created, the object is lifted into the air. At this point, the supporting columns, trees, tables and avatar people are created by the placement of prefabricated models at the desired locations. The building is then lowered by visual inspection onto the supporting columns. Next, the user performs further carving and a large hole is created through the centre of the building. After around 10 minutes for this example, the desired model is complete and the user can now move around the environment to preview it from different viewpoints via the see-through head-mounted display.

10.8 Augmented reality view of the new building (image reproduced in the colour plate section)

10.9 Virtual reality view of the new building

Conclusion

We believe that the use of mobile outdoor augmented reality systems will greatly enhance the ability of a planner to visualize new environmental and landscape designs. Key to helping the planner is the ability to view the designs in the field at the location of the changes. Being able to walk freely around the site allows the planner to *experience* the new designs and to better understand such attributes as colour, size, distances and interplay with existing structures.

This section has presented our existing system Tinmith as a prototype mobile outdoor augmented reality visualization system. The system allows users to place predefined graphical objects at the desired location, define planar regions of interest, and capture and create the shape of buildings. These tools allow the planner to both visualize the designs and modify them in the field to best suit the environment. Future technology advances will enable the construction of mobile AR systems that are lighter and more portable, for use in possible commercial applications.

VISUALIZATION AND PARTICIPATORY DECISION MAKING

Nathan H. Perkins and Steve Barnhart

Introduction

Participatory design and planning for large projects is often constrained by spatial and temporal limitations because involving large numbers of people in decision making is expensive and cumbersome. On many large projects, such as regional highway proposals, a significant amount of time and resources can be spent in scheduling and holding public meetings that are often not convenient, constructive or consistent with the often-stated purpose of involving the affected sector of the public in decision making. To be participatory, decision making requires removal of the barriers of limited access to information and the provision of more meaningful and descriptive information on the likely effects of decisions. Also, meaningful participation requires the creation of an efficient procedure for collecting and assessing responses to visualizations as necessary if decision makers are to respond appropriately.

Presenting commonly understood visual information on the consequences of decisions to potentially affected parties is a necessary early step in participatory design and planning. However, as sophisticated as visualizations have become with the advent of digital technologies, the process by which visualizations are used in communication – from presentation to discussion to decision – has remained quite traditional. With continuing advances in the representative quality of visualizations and, now common, access to widely available communication technologies, such as the Internet, opportunities for public participation in design and planning decisions need not follow tradition. It is our contention that meaningful participation in design and planning decision making is no longer *a priori* constrained by spatial or temporal barriers. In large part, the 'digitalization of visualization' has created a common language of producing and presenting (the 'Push') visual consequences of decisions and collecting and analysing responses (the 'Pull') to those potential consequences from a broad and often dispersed public. To maintain a balance in the Push–Pull relationship, both presenting and collected information need be planned for early in the design phase of projects, requiring stakeholder involvement if true participatory design is to be achieved (Figure 10.10).

Previous participatory efforts

Visualizing and presenting potential changes to the visual environment, based on design and planning decisions, has been practiced for centuries.

Prospects

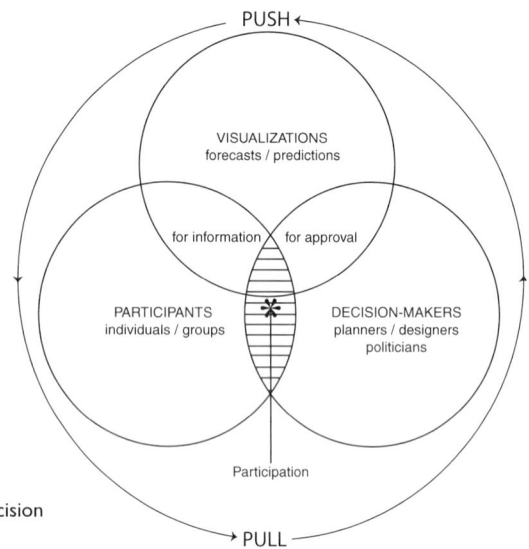

10.10 The push and pull of information in participatory decision making

The Red Books of Humphrey Repton (1816) utilized overlays bound in red leather to depict future conditions of landscape design based on proposed modifications to the existing environment. Repton's visualizations were used to great effect in conveying information to his clients but were of limited utility in conveying that information to a broad audience (see also Figure 1.4).

Pretty (1995) would probably refer to this type of participation as 'participation by consultation'. An expert presents a probable future condition arising from design decisions to a person or persons for review and for comment. In an ideal public process, comments are reviewed and reflected in revised proposals. Many projects, particularly those in the public realm, are required by law and/or regulatory mechanisms to have a minimum amount of public consultation. The scope, timing and level of consultation can vary from presentation alone, to collecting opinion, to involvement in the planning and design process and, in some cases, to the public's right of approval. This iterative process of planning and design is expensive and often unpredictable in its outcomes and, based on many professionals' experience, leads to the cynical conclusion that public involvement is a contentious and inefficient process.

In her classic paper, Arnstein (1969) proposed a typology of citizen participation organized by eight 'rungs of a participatory ladder' shown in Figure 10.11. The bottom rungs of the ladder are essentially non-participatory in that experts simply seek the illusion of participation while retaining authority. Examples abound where visualizations are created as an end product of an expert process and there is neither the capability nor intention to modify the proposed project. In some cases, the visualizations themselves are used as a marketing tool, pre-

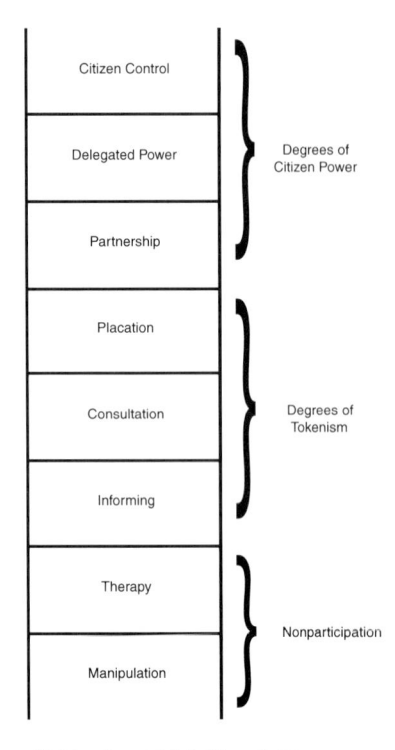

10.11 Arnstein's ladder of citizen participation

senting visual forecasts that lack even best-guess credibility, validity and reliability. Controversial projects often use visualizations to manipulate by using carefully selected viewpoints or 'artistic' licence to create pleasing yet inaccurate forecasts (Perkins and Barnhart 1996).

Climbing upward on Arstein's ladder, the middle rungs offer some degree of public involvement by informing, consulting and placating affected parties. Most project visualizations are likely to fall somewhere in this area. Conscientious and knowledgeable creators of visualizations attempt to meet the basic criteria of accuracy and realism proposed by Sheppard (1989) and have a working competency with the principles of visual analysis (Smardon *et al.* 1986). Meeting this level of participation suggests that, at a minimum, visualizations are intended to inform and that a process for collecting and assessing responses to those visualizations is in place. In many instances, regulations require that public consultation take place and public agencies find quality visualizations essential in presenting proposals and keeping communication among themselves and the public 'on track'. Our experience has shown, however, that the larger the proposed project is in scale, or the more complex the project, the more willing decision makers are to inform rather than engage the public on a higher rung of the participatory ladder. It may well be that larger, more complex projects simply have more information to push and with limited resources the push/pull equilibrium becomes unbalanced.

Meaningful participation, Arnstein argues, can only be achieved on the top rungs of the ladder that represent increasing degrees of citizen power such as partnership, delegation and finally control. Visualizations do not create this level of empowerment among citizens but they greatly facilitate the discussion. Visualizations, particularly those that are seen as drafts or aids in discussion towards decisions, can be used to mediate and ultimately lead to political empowerment as those people normally on the receiving end of decisions have the information needed to help form decisions (Rocha 1997).

The participatory approach

Traditionally, projects at a landscape scale such as highway proposals have been presented at a non-human-eye view such as two-dimensional plans, oblique aerial photography or schematic maps outlining project areas. This approach allows the designer to convey great amounts of information throughout the early stages of the project and often continues as the sole method for spawning a design dialogue with project participants. In each case the proposal is illustrated in fairly graphic manner, conveying the designer's concept and which can often lead to uncertainty in the minds of the non-professional audience as evidenced by frequently contentious discussion.

Even as digital tools become adopted in offices, their full power is usually under-utilized. What has become a deterrent to many design professionals is switching from the familiar approach of two-dimensional planning and presentation to incorporating three-dimensional and photo-realistic techniques in both the planning/design and presentation stages. Increasing levels of development of technologically sophisticated tools has provided the means by which three-dimensional information that surrounds complex topics such as landscape quality and natural environment preservation may be presented in a familiar and widely understood manner.

In the arena of planning and design what each of these varied levels of communication have in common is the central role that visualizations can assume as the expert-created stimulus in engaging the public in decision making through assessing their response to proposed change. There are still challenges to be overcome as this process continues to evolve. Three obstacles reoccur frequently to stand in the way of successful communication for the purpose of public input:

1. reaching the large silent majority of community interest through a broadened awareness of the issue at hand;
2. presenting the information in an engaging manner that is easily understood by laypersons and other non-design professionals; and
3. providing a convenient and efficient method of collecting large amounts information in a manner which is easy to analyse.

Developing a communication strategy that incorporates an information flow which is easily understood by all parties lays the critical foundation for fulfilling expectations that are both realistic and achievable.

Over the past decade we have sought to capitalize on the digital information revolution by creating applications that present visual information to the public *and* collect responses in a number of formats. The ultimate purpose of these projects has been to explore the communication process, hopefully leading to a better-informed public and a more efficient design and planning process. By combining visualizations with Internet-based participatory design, designers and planners may utilize a powerful and persuasive set of tools to engage large numbers of stakeholders in a manner never before possible.

The objective of an enhanced 'Internet-based' public forum for input should adhere to three key goals:

1. the web site *must* be promoted to attract a broad audience and communicate the project issues in a manner easily understood by laypersons;
2. the web site should become a vehicle that will allow for more effective messaging, such as opportunities for online discussion or submission of comments; and

3 the web site should foster an increased sense of community through the education and appreciation of the environmental issues proposed. The following two case studies provide examples of this approach.

Case study: arterial route selection Cambridge, Ontario

Traffic congestion within the city of Cambridge, Ontario (population 113 000) led regional planners to propose a new traffic corridor within the rural lands surrounding the city core. As part of a class environmental assessment, the visual quality of the lands within the study area was required to be evaluated in the course of the planning process. The visual quality of the landscape was evaluated and then categorized and presented on an interactive Internet site for public participation. An online visual survey instrument provided a means to test the sampled landscapes for visual preference by allowing respondents to choose their preferred landscape scenes. The remote data collection occurred over several months as residents connected online to complete the survey and final results were used to create a reference map that identifies areas by degree of visual quality.

The objectives of the study were to:

1 develop a reliable method of identifying and producing an inventory of areas of visual significance;
2 identify areas that should be preserved for their visual quality and areas which are more suited to development;
3 conduct visual preference testing of landscapes as a means for providing public input on landscapes within the study area; and, last,
4 create a reference map that identifies landscapes by degree of visual quality.

A photographic survey was conducted on a 500–1000 m grid within the study area using the existing road network for reference. A total of 153 photographs were taken and categorized into groups of high, medium and low visual quality. An Internet survey was established to provide opportunity for the general public to rate landscape scenes on the basis of visual significance (a preference rating for a given scene based on existing conditions) and visual sensitivity (insight into future conditions, 'if some form of development were to occur which scene should be preserved exactly as is?'). The survey ran for six months on a dedicated Internet server allowing for the automatic collection and tabulation of results. The images provided a means for laypersons to understand the concepts of visual sensitivity and visual significance and provide their comments on these visually descriptive concepts.

Prospects

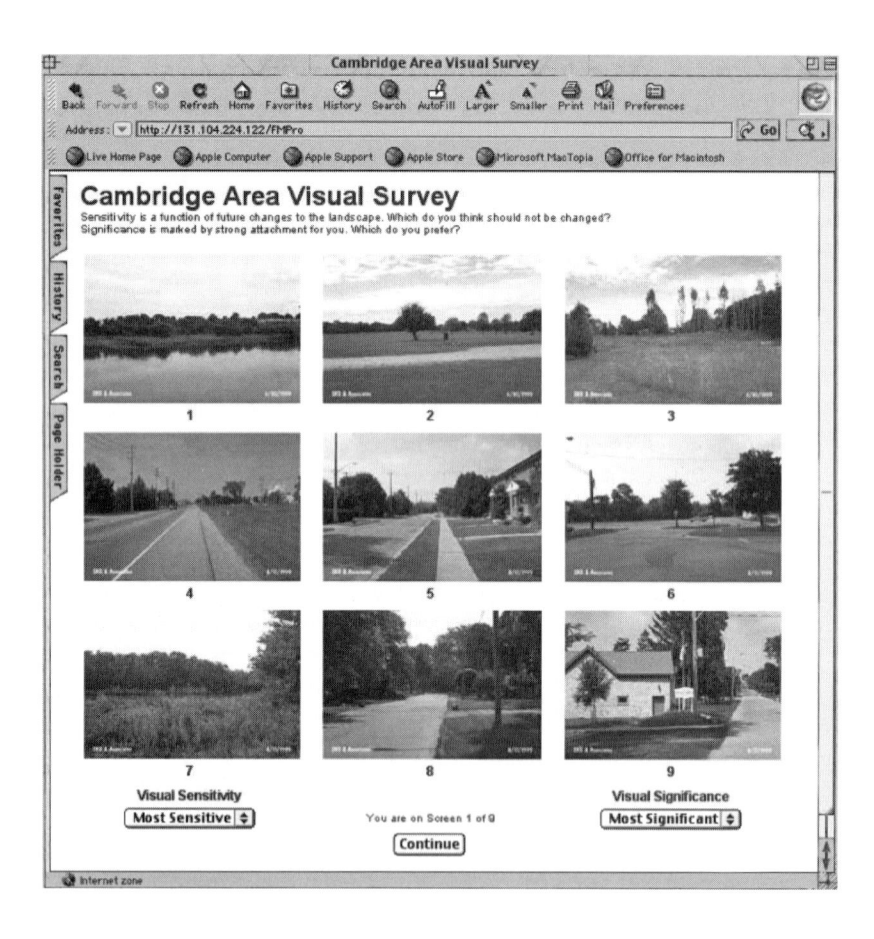

10.12 Cambridge Survey page

The landscape visual inventory provided the background information to aid in the selection of alignment alternatives and to help determine which locations would require visual mitigation measures. The online visual survey provided the general public with a means of reviewing landscapes interactively and stating their preference for significant and sensitive lands. The Internet survey became an effective tool for viewing and collecting public input on complex information such as visual landscape in a manner easily understood (Figure 10.12).

Case study: Red Hill Creek, Hamilton, Ontario

The Red Hill Creek Valley is a 650 ha (1600 acre) park and open space area serving a community of over a million residents. It also forms one of the few remaining natural links between the Niagara Escarpment (a UNESCO Biosphere Reserve) and Lake Ontario. Biological inventories have shown it to contain 24 species of mammals, 92 species of breeding birds, 18 species of fish, 47 species of butterflies and 571 species of plants. This biodiverse area provides green space for the east end of the city of Hamilton, a heavily industrialized urban metropolis. However, existing development has degraded the performance of Red Hill Creek, which is causing serious erosion problems in the valley.

The Red Hill Valley is the site of a proposed five-mile expressway intended to divert heavy truck traffic off local city streets. The expressway, which has been under study since the 1950s, and was approved in the 1980s, will connect an existing expressway located above the Niagara Escarpment through the valley to the Queen Elizabeth Highway along the shore of Lake Ontario. Since 1997, the expressway design has been refined to reduce impacts to the valley. The design now includes a 250 yard viaduct structure at the escarpment designed to preserve trail and wildlife linkages. Approximately 4.4 miles of the creek will be reconstructed using a 'natural channel' approach to reduce erosion, restore floodplain functions and minimize channel crossings by the new road. Habitat restoration measures elsewhere in the watershed will help to compensate for the loss of habitat and functions in the valley.

As part of the fulfilment of requirements of an Environmental Assessment, a public communication was to be developed and maintained by the local municipality responsible for the development of the expressway. In 2001, SKB & Associates was retained by the City of Hamilton to conduct an extensive Visual Impact Analysis through computer simulations to provide the residents and Red Hill Valley project design team with a thorough understanding of the anticipated effects of the proposed expressway. Through field investigation and photographic analysis, representative visual impact locations were recorded to provide the framework for a comprehensive virtual tour of the proposed expressway. The final product reflects the evolution of several months of design development and comments from project partners.

During the development stage, while web site access was restricted to only municipal staff and project partners, the computer simulations became a tool to fine tune design details. By employing rich media (including maps, graphics, photographic images, three-dimensional models, impact mock-ups and textual support materials) the project realized its special value as an interactive communication tool. The interactivity, paired with three-dimensional visuals, encouraged the members participating in the study to share ideas, concepts and reactions to the study results. By lifting temporal and spatial boundaries between individuals participating in the exchange, the web site medium formed the nexus of the communication network.

Building on the experience gained during the development stage, the goal of web site design became one of providing a unique forum to allow residents adjacent to the project and the public at large to explore proposed change. An objective of the authors was to create an engaging media that would stimulate and solicit public input.

Prospects

At the outset of the project a data-driven secure web site was deployed to minimize travel time and coordination of project meetings while providing an ongoing working paper that informed the client members of current tasks and responsibilities. This initial presentation allowed all parties to fully understand the scope, expectations and activities inherent in the project. It provided a forum for interaction between the participants and allowed for 'instant' presentation of whiteboard ideas that could spark a creative exchange. As the project progressed, additional components were added to the web site, leading toward a full working prototype.

Technical design focused on the range and type of Internet users that may access the site. Graphically the imagery would provide a photo-realistic interpretation for the viewing audience. Computer animation although engaging, was technically cost-prohibitive due to the requirement of recreating the look of the natural setting. Through perspective matching techniques, highly realistic simulations were created as still and interactive panoramic movies. The web site was deployed on a UNIX server and took advantage of Apple's QuickTime standard for Virtual Reality presentation of photo-realistic images.

The web site design approach employed technologies that would not limit its use by the widest possible audience. Every effort was made not to include elements that relied on proprietary components not found on most computers. During web site development, the value of enhanced web site interactivity was weighed against the potential for such technology to constrain less sophisticated computer users. Consequently a JavaScript driven interactive process was judged to be the most appropriate method to maximize user experience and user audience.

The final site (www.skbandassociates.com/redhill) provides for a full discussion of the nature of Visual Impact Assessment, the project scope and a visual tour of the proposed project. On entering the web site, the user is greeted with a general description of the site contents. Navigational choices are in graphic format. The main navigational tool is a full view map of the proposed expressway project. It provides an overview of the scope and also allows for user interaction by 'mousing' over different parts of the valley and the expressway for more specific visual impact evidence. The user has the opportunity to explore the impact of the proposed expressway project through either a guided tour from one vantage point to the next or at randomly selected vantage points from the map (Figure 10.13).

Individual impact point screens provide a detailed context map of where the view location is relative to the proposed project. Each location contains a smaller overview map showing the surrounding context in two-dimensional plan form as well as photographs illustrating the area before the proposed changes and a photo-realistic model of the proposed the changes (Figure 10.14).

In the cases where extensive visual change is proposed, still images and complete 360° panorama views of the visual impact were included.

Visualization prospects

A total of 29 viewpoints were selected to represent the potential key points of visual impact. Each point was photographed in a variety of seasonal settings and mapped to establish its context relative to the proposed expressway. GPS positioning provided a reference for the perspective matching approach.

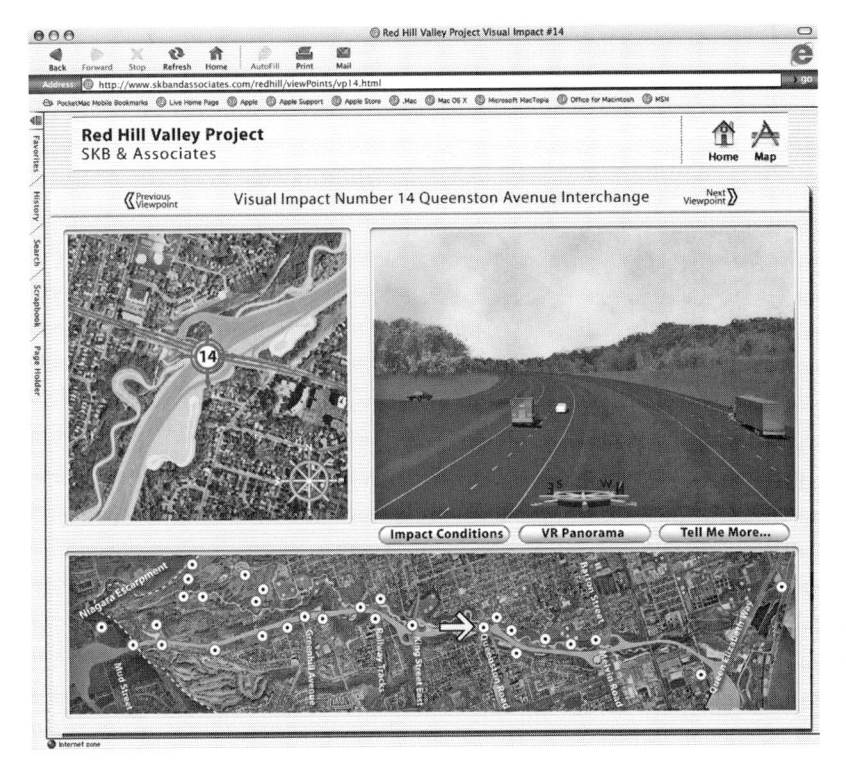

10.13 Red Hill Valley Main Navigation Page showing from top left and then clockwise: individual view locations with navigation arrows, 360° panorama and overall key map (image reproduced in the colour plate section)

10.14 Existing condition (left) and simulated proposed condition (right)

249

The virtual tour web site was designed to provide interested members of the public with an understanding of proposed changes to the Red Hill Creek Valley. The project also aided in the development of detail design. The Internet site has been incorporated into the City of Hamilton's web site. To aid city staff and the project team in discussions with municipal golf course members concerned about impacts to the course the Internet site was later expanded. Views of the proposed changes around the course were used to demonstrate the extent of change which ultimately alleviated members' concerns.

Conclusion

With the continuing evolution of digital technologies, there is no reason to expect that the tools and techniques of visualization will not continue to become more sophisticated, less costly and more pervasive in planning and design at all scales. Visualization is, however, only half of the process necessary to develop meaningful communication among experts and the public. Visualization, no matter how sophisticated, is a 'push' technology akin to a monologue. When the public is asked to participate, or public assessment and commentary are required, the balancing 'pull' leads to a participatory conversation. Many of the techniques described in the case studies above can be applied in differing contexts and across many scales. Presenting visual information in a digital format via the Internet has provided many immediate benefits, not least of which are substantial savings in time and resources.

In large complex projects with a large number of people involved, having a current record of the project has also facilitated project member workflow and communication. By presenting realistic and engaging visual information accessible at anytime from any location with Internet access, the public benefits from improved communication leading to increased levels of project participation. As many project coordinators realize, an informed public can be an asset and a poorly informed public can often lead to a contentious and difficult decision making process.

What remains uncertain at this point is how effective has this approach been in comparison to traditional procedures? Clearly, research needs to be conducted on a range of issues that would allow direct comparisons to be made. At present, however, emphasizing both visualization and participation as integral components in design and planning has been successful in the Canadian context.

VISUALIZATION IN SUPPORT OF PUBLIC PARTICIPATION

Michael Kwartler

Introduction

Our underlying assumption is that there will be growth, with development pressures on greenfields, typically undeveloped or agricultural land. The impetus to look for improved planning methods is a result of, and negative reaction to, current development practice. In Baltimore, for example, we have seen automobile-based planning manifesting itself in traffic jams, chaotic development patterns and the loss of open space.

Each planning situation is different in scale, landscape, degree of anticipated change, governmental and regulatory structures, and the nature of public participation. The integration of simulations and visualizations into the public participation process, to facilitate informed discussion and decision making, can be guided by the people involved, propinquity and localism. We have used technology consistently to develop plans in a variety of settings from hands-on workshops to large 'town hall' meetings (Figure 10.15). In our work, the technology was integrated into the process and used at the front-end to help formulate goals and the vision. Public expectations about the use of the

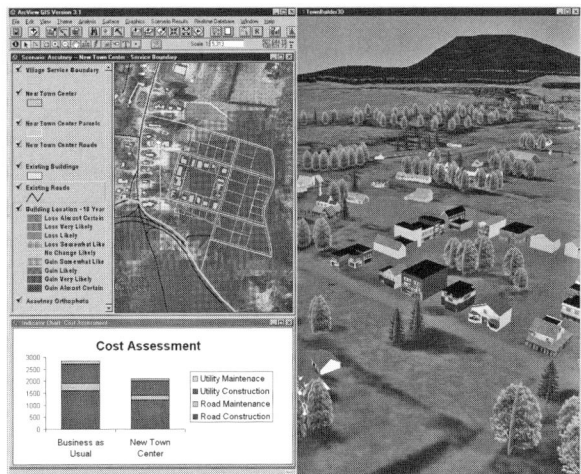

10.15 Systems for public participation: (left) scene from a hands-on workshop with Steering Group in Santa Fe; (right) software for option evaluation combining mapping, three-dimensional views and outcome summaries

visual simulations, on the other hand, have typically been limited to use at the back-end to sell the plan.

Our recent studies have commonalities that are useful for comparing why, how and under what circumstances visual simulations were used to enhance public participation and decision making. Conceptually, they share a common base in visioning. Visioning is a citizen-driven process where the results are derived from public input. Its purpose is to reach consensus on issues regarding values and group identity. The process is inclusive, neither top-down nor bottom-up, and includes stakeholders (from business leaders to NGOs) and the public sector.

Place plays a critical role in locating common ground. As Donald Appelyard (1979: 143) observed: 'technical planning and environmental decisions are not only value based ... but identity based'. From the perspective of the lay public, place (a neighbourhood, town, city or region) is experienced as a whole. The quality of place, the combination of its experiential and functional attributes and group values and identity, is fundamental to visioning. Visioning uses physical design as a form of inquiry, exploring the match and mismatch between words and images; abstractions that have very specific real world implications. In this process, it is not unusual to hear the shock of 'that's not what I meant at all' when words and numbers in a standard master plan or zoning resolution are simulated and visualized dynamically in three dimensions. The visual simulations used in visioning play a similar role; grounding metaphor in reality.

Baltimore: Vision 2030 – shaping our region's future together

Over a 15-month period, the Vision 2030 public Thematic Committees explored six thematic areas that are brought together to form a regional perspective (Baltimore Regional Transportation Board, 2003). The areas were:

- economic development;
- education;
- environment;
- government and public policy;
- liveable communities; and
- transportation.

The visioning process involved six interrelated and sequential steps:

- Step One: understanding the region – perception and reality.
- Step Two: involving stakeholders (the Regional Workshop).
- Step Three: prototypical development patterns and scenarios.

Visualization prospects

Table 10.1 The tools used and applications developed to seamlessly integrate words, numbers, maps and images in a real-time three-dimensional environment

ArcView GIS	Geographic Information System (GIS) developed by Environmental Systems Research Institute (ESRI) and includes 3D Analyst and Spatial Analyst
CommunityViz	Suite of ArcView based decision support software designed by the ESC for the Orton Family Foundation that includes: a 'scenario constructor' module for developing alternative scenarios and assessing their impacts; a 'policy simulator' module to predict possible futures; and an interactive 3D three-dimensional module where alternative scenarios can be designed and visualized and where citizens can change the scenarios in real time.
Creator	3D Three-dimensional modelling and authoring software for real-time simulation and visualization developed by Multi-Gen Paradigm;
TerraTool	Parametric modelling software for real-time viewing developed by TerraSim, Inc.

- Step Four: gathering ideas and testing results with the public (the Regional Public Meetings).
- Step Five: developing vision statements and strategies.
- Step Six: testing the vision statements and strategies with the public.

In the Vision 2030 process, visual simulation played a central role in helping the public and the Thematic Committees reach consensus on the 'hot button' issue of where and how to accommodate growth in the region identified by the focus groups in Step One. The software used in the workshops is introduced in Table 10.1.

The Regional Workshop ('where to grow')

In response to the 'hot button' growth issue, the Regional Workshop focused on 'where to grow'. Organized as a game, the purpose of the workshop was threefold: to understand the complexity of thinking regionally; to gain 'intuitive' public input on future growth and land consumption considerations; and to prepare for future subcommittee work (e.g. developing the vision statements, strategies, and principles that formed the core of Vision 2030).

The participants included 65 stakeholders: elected officials, planners, educators, citizen activists, staff from NGOs, and business leaders. Participants were evenly divided into eight groups, each with a facilitator.

Participants agreed on a percentage (average of all eight groups) of the region's land they wanted to protect over 30 years, in addition to land already protected. The next step was to agree on a common set of criteria, weighted differently by each of the eight groups, used to select the areas to protect (e.g. creation of contiguous natural environments, protecting forest and trail areas, etc.).

Prospects

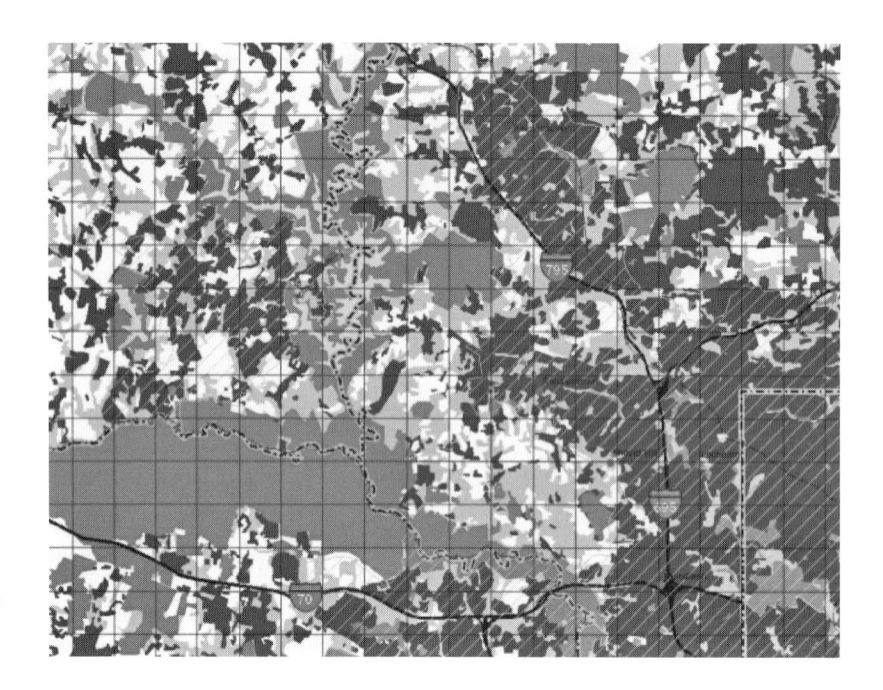

10.16 Detail of map of the Baltimore Region with mile square grid and GIS coverages

A large GIS-generated map of the region, which included layers delineating urbanized areas, protected areas, and agricultural and other unprotected land, was overlaid by a 2.56 km^2 (1 mile^2) grid (Figure 10.16). Each group was given green 'chips', each representing 1 grid square of land and asked to place them onto areas that the group believed should be protected. The results of each group's approach to future land protection were tabulated and workshop average calculated. The patterns of chip placement were compared and discussed by the workshop participants, revealing an underlying consistency in the choices.

The next step was to locate growth. The groups were given brown 'chips' that represented the amount of land that would be needed to accommodate the region's projected growth for the next 25 years. The participants agreed on criteria to guide their decision making for allocating growth (e.g. along transit lines in already developed areas, in undeveloped areas, near employment centres, etc.). The emerging consensus was to locate growth in the region's developed areas and protect undeveloped and agricultural land.

The Regional Public Meetings ('how to grow')

Over a two-month period, 17 facilitated Regional Public Meetings were held. Presentations were made of prototypical development patterns, region-wide development scenarios, and the absolute and relative performance of the development scenarios, questionnaires administered, and small group idea sessions were conducted.

The Thematic Subcommittees identified three development patterns, and four future regional development scenarios whose relative performance would be evaluated by agreed indictors. The development patterns were modelled in three dimensions as building blocks. Each scenario showed how the region would develop depending on the allocation of the development patterns.

The three development patterns reflected trends occurring in the Baltimore region as well as nationwide. Each had different implications for land consumption, mix of housing types and proximity to jobs, shopping and entertainment. They were as follows.

- Type A: Conventional development pattern in undeveloped land. It reflects a continuation of how the region has been growing with single family detached houses, and shopping, entertainment and employment in auto-centred malls (Figure 10.17(a)).
- Type B: Mixed-use walkable community on undeveloped land. It assumes the creation of more compact neighbourhoods with a mix of housing types with shopping, entertainment and employment nearby (Figure 10.17(b)).
- Type C: Mixed-use walkable communities on redeveloped land. It also assumes the creation of more walkable compact communities but on redeveloped land (Figure 10.17(c)).

Each building block or development pattern had the same brief or programme; accommodate 1000 households – with supporting commercial resources, schools and open space. This allowed for 'apples-to-apples' comparisons.

Vision 2030 was not focused on a particular place in the Baltimore region but rather on characteristic areas in the region. Given the region's emphasis on the local control land-use regulation), it was critical that Vision 2030 not appear to be usurping local authority (Figure 10.18). Working collaboratively with Thematic Subcommittees, two representative places were composited using the five county GIS orthophotographs and database; one with large tracts of undeveloped land and existing suburban and rural development patterns with the possibility of 'greenfields' development and the other an urbanized centre with the possibility of infill redevelopment.

Using the briefs, prototypical building types and development patterns were designed (layout of blocks, lots, streets, uses, open spaces, distribution of building types, etc.) modelled in three dimensions and inserted into the 3D/GIS. The real-time 3D/GIS environment allowed for an efficient confidence building process where the Thematic Subcommittees viewed patterns dynamically, commented on the design of the development pattern, iteratively refined it and selected views and real-time walk-through paths to be presented at the 17 Regional Public Meetings.

The three-dimensional visual simulations of the three development patterns were then assembled into the four regional development scenarios:

Prospects

10.17 (a) Conventional development 1990–99 trends in the region; (b) mixed use – greenfield development; (c) mixed use – redevelopment

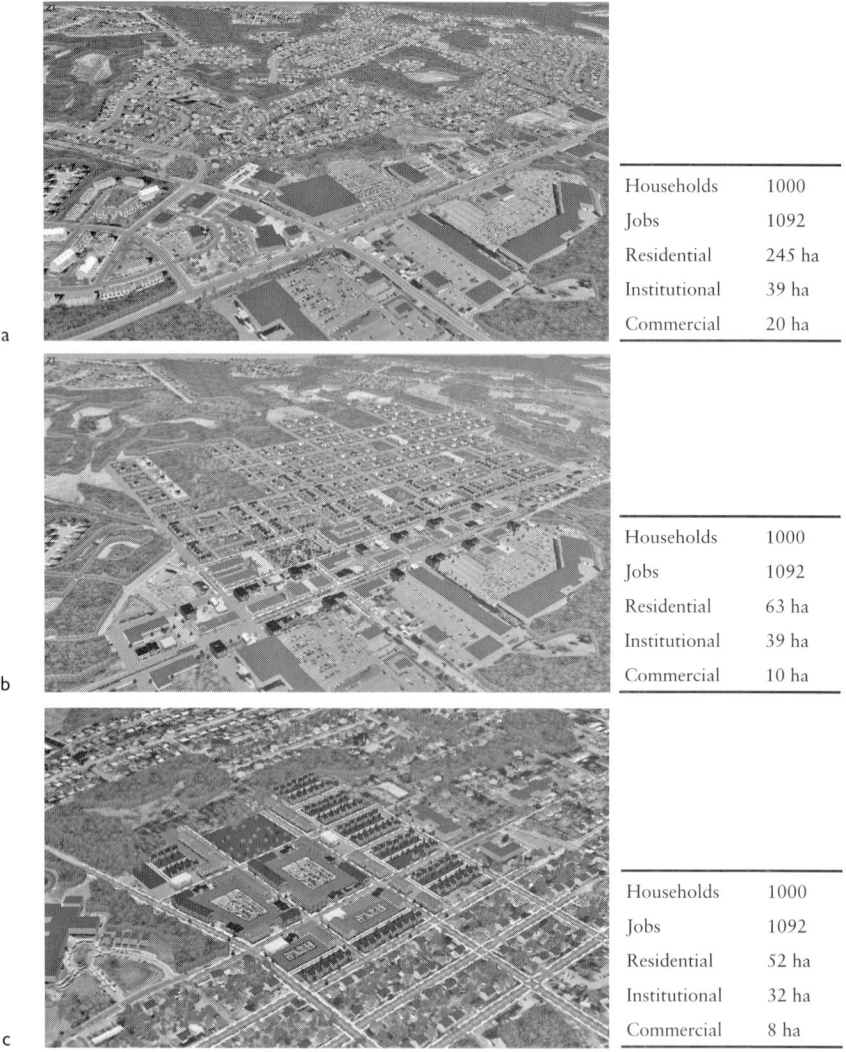

a

Households	1000
Jobs	1092
Residential	245 ha
Institutional	39 ha
Commercial	20 ha

b

Households	1000
Jobs	1092
Residential	63 ha
Institutional	39 ha
Commercial	10 ha

c

Households	1000
Jobs	1092
Residential	52 ha
Institutional	32 ha
Commercial	8 ha

1. current trends and plans;
2. emphasis on road capacity;
3. emphasis on mass transit; and
4. emphasis on redevelopment.

The scenarios accommodated the forecasted population and employment growth for the region by using the development pattern types in different combinations. The compositing of development pattern types or building blocks into scenarios is illustrated by comparing the mix of development types between scenario 2 and scenario 3 (Table 10.2).

Visualization prospects

10.18 'Greenfields' composite area

Table 10.2 Two of the Baltimore region development scenarios

Scenario 2 Emphasis on road capacity	
Type A: Conventional development pattern on undeveloped land	75%
Type B: Mixed-use walkable communities on undeveloped land	20%
Type C: Mixed-use walkable communities on redeveloped land	5%
Scenario 3 Emphasis on mass transit	
Type A: Conventional development pattern on undeveloped land	25.0%
Type B: Mixed-use walkable communities on undeveloped land	37.5%
Type C: Mixed-use walkable communities on redeveloped land	37.5%

The comparison of the four scenarios was measured through a computer model that used ten indicators to measure the performance of each scenario. They included, for example, acres of new land consumed by development from the year 2000 to the year 2030, percentage of new neighbourhoods that provide choice of housing types and range of prices, air pollution from vehicles, impact of existing and future development on existing water quality in Chesapeake Bay, and percentage of new jobs accessible by transit. The data used by the computer model came directly from the 3D/GIS visual simulations of the development pattern types.

Prospects

10.19 Eye-level walk throughs: (left) existing condition; (right) redevelopment (images reproduced in the colour plate section)

The three development patterns, the four regional development scenarios, and each scenarios' performance were presented at each of the 17 Regional Public Meetings in conjunction with a questionnaire entitled 'Choices for the Future'.

The scenarios and performance indicators represent 'what ifs', hypothetical situations that were intentionally designed to offer a wide range of choices. Their abstraction, particularly when expanded to the five county region, was made palpable by the visual simulations that employed three-dimensional models and eye-level walk-throughs (Figure 10.19). They communicated to the public 'if you think or prefer this, this is probably the kind of place that will result, and is it acceptable?' The visual simulations were compelling and, in conjunction with the performance evaluation of each scenario, provided a comfort level for respondents to the 'Choices for the Future' questionnaire to overwhelmingly support the redevelopment and emphasis on mass transit scenarios, both of which consumed less than half the amount of land in comparison with scenario 1 – current trends.

The way in which visual simulations were used in Vision 2030 was a function of the scale of community participation process in the region. The Thematic Committees were hands-on using the real-time environment provided by the 3D/GIS. Similarly, the Regional Workshop allowed for direct interaction with the 3D/GIS and the visual simulations. The results of the 'where to grow' and 'how to grow' workshops were presented at the 17 Regional Public Meetings which, due to the large number of participants, could not be hands-on in terms of determining or even changing the content. Rather, the Regional Public Meetings were designed to elicit community response to a series of scenarios. The public outreach and communications effort, the visual simulations, the indicators used to evaluate the performance of the development patterns and scenarios, and the iterative nature of the process were very effective in keeping people informed and involved and in communicating the ideas developed in the committees and Regional Workshops. Telephone interviews of 1200 randomly selected participants carried out at the end of the visioning process validated the positions taken by Vision 2030.

Conclusion

Because people experience the world in three dimensions, in time, and in motion, digital technologies that mimic human experience are most easily understood and accepted by the public during the planning process. The tools used were able to seamlessly integrate words, numbers, maps and images in a real-time three-dimensional environment supporting place-based planning. They supported public participation in the development, review and refinement of the three-dimensional building blocks and the compositing of alternative scenarios. They also built public confidence in the visual simulations as a result of being able to use the 3D/GIS interactively, and the performance evaluation of 'what if' scenarios.

A clear example of this emerged in a case study based in Santa Fe, NM. Consensus was reached on a housing density of over 12 DUs/ha (5 DUs/acre) or almost twice the current zoned density. The real-time 3D/GIS models went beyond abstractions such as FAR (Floor:Area Ratio – a poor indicator of density and/or intensity of use found in most zoning regulations) providing both the quantitative effects of sprawl and urban densities on land consumption and the experiential basis to discuss in concrete terms the kind of place the residents and stakeholders wanted. Without having first developed the vision for South-west Santa Fe through simulation and visualization, it is highly unlikely that the community would have considered doubling the density, far less supported it.

Visual simulations support both deductive and inductive reasoning (analysis/synthesis) when their applications are fully integrated with the design of the planning process. Real-time 3D/GIS was critical in supporting a planning process that was engaging, informative and ultimately instrumental in helping communities to reach consensus about their future.

Acknowledgements

Vision 2030: Shaping our Region's Future Together was an initiative of the Baltimore Regional Transportation Board, the federally mandated Metropolitan Planning Organization for the Baltimore region. Additional support was provided by two NGOs, the Baltimore Metropolitan Council, a private non-profit-making regional planning agency and the Baltimore Regional Partnership, an alliance of civic and environmental groups that share a common agenda of enhancing quality of life through community revitalization and environmental protection. ACP – Visioning and Planning (ACP) with its expertise in conducting regional visioning was the lead consultant. The Environmental Simulation Center (ESC) was a member of the ACP team of five sub-consultants.

TRENDS, CHALLENGES AND A GLIMPSE OF THE FUTURE

Ian D. Bishop and Eckart Lange

Trends and challenges

Visualization is at the brink of a revolution in application potential as a result of the sudden emergence of high performance desktop computer graphics, large volumes of digital spatial data, increasing Internet bandwidth, enhanced positioning technology, easier systems integration and widespread demand for a public role in environmental decision making.

Desktop graphics

Desktop graphics were reviewed in detail in Chapter 4, 'Visualization technology', p. 51. It is worth reiterating here that whereas even two years ago few people bought computers with a specialized graphics card, today they are virtually standard equipment – especially in households where computer games are one of the major motivators. This means that most users can explore complex three-dimensional models interactively. Such graphics cards will next find their way into pocket-sized computing devices (including mobile phones) and, as such, will be able to support augmented reality applications with minimum weight, intrusion and inconvenience.

Spatial data infrastructures

Today everyone is talking not simply about digital data but more pervasive spatial data infrastructures (SDI). Data are increasingly coordinated and made accessible online. Specialized data definition systems based on standard protocols defined by the Open GIS Consortium (OGC) will allow a user to access multiple datasets from multiple sites without explicit knowledge of their source. The talk now is about adding a layer of widely accessible generic tools between the data and the user in order to allow individual value-adding to the data (Feeney *et al.* 2002). The next step is a movement towards automated development and either display or distribution of three-dimensional models from online SDI.

Positioning and orientation

As Piekarski and Thomas (see 'Future use of augmented reality for environmental and landscape planners', p. 234) pointed out, the key to augmented reality application is effective information about one's position and orientation. Since the first positioning satellites were launched in the 1960s, determining position has become easier and cheaper. Using global positioning system (GPS) technology, determination of location with an accuracy of around 1 cm to 15 m is possible, depending on the mode and type of GPS equipment used. Real-time kinematic (RTK) GPS, based on receipt of not only the satellite signal but also a signal from a known ground location, can provide an accuracy of 2 cm or better.

The use of multiple GPS receivers can, in theory, provide both position and head orientation. However, there are practical impediments to personal use related to signal interruptions and achievement of sufficient spacing between GPS units. Coming to the aid of GPS for continuous position and orientation information is a range of new, low-weight devices for inertial and dead reckoning (DR). Inertial and DR tracking systems are self-contained instruments that provide relative positioning solutions. The relatively new technology of MicroElectro Mechanical Systems (MEMS) allows for a tighter, more compact integration of sensors with a complete inertial navigation system existing on a single microelectronic chip (Kourepenis *et al.* 1998).

Applications of the type described by Piekarski and Thomas thus become increasingly viable. The potential then exists for people to share virtual space – with all parties in augmented space, some in the augmented space and others in a corresponding wholly virtual world or (as mentioned above) all in the virtual world.

Systems linkages

Communications between far-flung computers depends upon telecommunications developments: broadband Internet, fast wireless connection and use of mobile phone technology. Broadband allows people to download complex three-dimensional models in a reasonable time. In the world of computer gaming, people can already fight, or better collaborate, with each other through the web. Mobile phone technologies will allow people in the field to join their office-based colleagues in the same virtual worlds.

At the same time, systems integration, especially using existing software packages and widely recognized standards and protocols, seems likely to accelerate. To interoperate properly, systems must have a common understanding of the data and operations they share. In the GIS context, the information communities' model of the OpenGIS consortium (OpenGIS 1998) addresses the problem of common understanding. Extensions to three-dimensional models seem inevitable. In addition to being able to communicate with each other, systems have to understand each other. The characteristics of a road and the language to describe it need to be the same from the two-dimensional data sources to the three-dimensional models. Researchers are addressing this issue of data vocabulary and calling it ontology.

Public demand

Contemporary users of spatial information technology have been privileged to work in a period when wonderful new tools are an annual event. As the technology has moved rapidly forward we developed applications without always paying great attention to their effectiveness. Those who develop, or seek to apply, advanced public participation systems need greater awareness of best practice in application. At the moment this is largely a matter of trial and error. We tend to repeat each other's successes and failures. We need more researchers (like Perkins, Barnhart, Kwartler and others in this volume) willing to analyse deeply their experiences in public interface with emerging systems.

Everyone has an interest in their own environment. Thus, visualization technology is increasingly being applied in selling, opposing and adjudicating developments and environmental management practices. People in many disciplines design, plan for or manage natural or built environments. The professionals, their clients and the people affected by their decisions should understand how the technology can be used and misused.

To conclude this volume we have prepared a scenario of future decision making in the landscape which makes use of the technologies described in the preceding chapters and the trends identified above. The remarkable thing about this scenario is that it could be happening now. The technology is already available and the software cannot be far behind. The stumbling point may be the will to put the possibilities into practice.

A glimpse of the future

A group of local residents are moving around on a low hill. Several higher areas are clearly visible across the adjacent valley. They are wearing half-silvered glasses linked to pocket computers which have in-built GPS devices. The glasses are equipped with MEMS-based inertial positioning. Several architects in nearby towns have their own computers showing a virtual view from the hill. The model has been downloaded to their computers from the national SDI after a quick selection on the Spatial Data Authority web site. Most are looking at their usual wide panel displays, one or two have donned glasses incorporating immersive display technology. In another country, on the far side of the ocean, a small team of engineers are in an immersive projection facility.

The people on the hill are the future residents of the area – they have bought blocks of land in the new subdivision, the architects are in the process of designing houses for these people and the engineers are in charge of a wind energy project to be installed nearby. The future residents have chosen to come to the site in person, even though they too could have seen the proposals remotely, because they want to be sure about the visual impacts of the wind farm development. They know that distortions are possible in other display contexts.

The engineers have undertaken analysis to determine the optimal location (in cost-benefit terms) for each of the 24 turbines. The residents can see these in their AR displays as they move around. Most have been directed to their chosen house lot via their GPS units. Preliminary architectural designs have already been prepared for most of the houses and these are also visible in the AR display. Many of the residents-to-be are standing at the location of either their kitchen window or front door depending on the orientation of the house relative to the proposed turbines. The architects have their CAD software integrated with the virtual environment display so that they can rapidly adapt the design if necessary.

A number of the residents signal through head movements that they not happy with the current arrangements. The visual exposure to the turbines is too high even with the typical atmospheric conditions for the area included in the display. Using tracked hand movements they point out the turbines that they feel are particularly problematic. These indications are fed to the engineering team who find that the majority of the residents' concern is focused on just four of the turbines. Perhaps these can be relocated without significant extra cost or loss of energy output. Activating once more their GIS-based windfarm layout analyser, they run a visual impact analysis which indicates that the turbines are in a potentially sensitive zone and also shows areas of lower sensitivity which can still produce acceptable energy outputs. They drag the four towers to their new proposed locations.

The residents acknowledge that this is an improvement. Several are no longer concerned as long as the towers are not brought back into a sensitive zone. Two residents are still unhappy with two of the turbines. The engineers indicate that they cannot find a site for these turbines that they are willing to accept. They bring the relevant architects into the discussion. Each of the two architects involved can clearly see the situation and they turn to their CAD systems to seek a design which will minimize exposure for someone inside the house or someone using the major entries. In one case a minor reorientation achieves the desired result. In the other case no solution is found. After some discussion (which is masked from the other users) between the resident and the engineers, a package of compensation involving provision of mature trees for the garden (these are shown as texture maps from the tree library on the resident's display) and some other measures (which are commercial-in-confidence) an agreement is reached. A final check is made under clear sky conditions and in this case no increase in concern is registered (this is quite rare, typically further adjustments are necessary at this point).

Agreement is signalled to all parties. The residents are comfortable with the arrangements, the architects confirm the preliminary designs and remind themselves to start work on the design details in a month or so and the engineers open the office bar (it being evening in their part of the world) and celebrate having survived the public consultation with minimum cost to the project.

REFERENCES

Preface

Ervin, S. and Hasbrouck, H. (2001) *Landscape Modeling: digital techniques for landscape visualization*, New York: McGraw-Hill.

Chapter 1

Allison, R.S., Harris, L.R., Hogue, A., Jasiobedzka, U., Jenkin, H., Jenkin, M., Jaekl, P., Laurence, J., Pentile, G., Redlick, F., Zacher, J. and Zikovitz, D. (2002) Simulating self motion II: A virtual reality tricycle. *Virtual Reality*, 6: 86–95.

Appleton, K.J. and Lovett, A.A. (2003) Defining 'sufficient' realism in visualisations for rural environmental decision-making. *Landscape and Urban Planning*, 62: 117–131.

Appleton, K.J., Lovett, A.A., Sunnenberg, G. and Dockerty, T. (2002) Visualising rural landscapes from GIS databases: a comparison of approaches, possibilities and problems. Computers, *Environment and Urban Systems*, 26: 141–162.

Appleyard, D. and Craik, K.H. (1978) The Berkeley environmental simulation laboratory and its research program. *International Review of Applied Psychology*, 27: 53–55.

Asisi, Y. (2004) 8848 Everest das größte Panorama der Welt (www.8848everest.de).

Azuma, R., Baillot, Y., Reinhold, B., Feiner, S., Julier, S. and MacIntyre, B. (2001) Recent Advances in Augmented Reality. *IEEE Computer Graphics and Applications*, 21: 34–47.

Beck, M. (2003) Real-time visualization of big 3D city models. *International Archives of the Photogrammetriy, Remote Sensing and Spatial Information Sciences*, Vol. XXXIV-5/W10.

Beck, M. and Steidler, F. (2001) *CyberCity Modeler und TerrainView – Werkzeuge zur Visualisierung von 3D-Stadt- und Werksmodellen*. VPK Vermessung, Photogrammetrie, Kulturtechnik, 7, 472–476.

Bergen, R.D., Ulricht, C.A., Fridley, J.L. and Ganter, M.A. (1995) The validity of computer-generated graphic images of forest landscape. *Journal of Environmental Psychology*, 15: 135–146.

Beveridge, C.E. and Schuyler, D. (1983) *The Papers of Frederick Law Olmsted*. Vol. 3, Creating Central Park, 1857–1861. Johns Hopkins University Press, Baltimore, London, 470 pp.

Bishop, I.D. and Dave, B. (2001) Beyond the moving camera: systems development for interactive immersive exploration of urban environments. Computers in Urban Planning and Urban Management, Honolulu, July 18–21.

Bishop, I.D. and Leahy, P.N.A. (1989) Assessing the visual impact of development proposals: the validity of computer simulations. *Landscape Journal*, 8: 92–100.

Bishop, I. D. and Rohrmann, B. (2003). Subjective responses to simulated and real environments: a comparison. *Landscape and Urban Planning*, 65: 261–277.

Bishop, I.D., Wherrett, J.R. and Miller, D. (2001) Assessment of path choices on a country walk using a virtual environment. *Landscape and Urban Planning*, 52: 227–239.

Bosselmann, P. (1983) Visual impact assessment at Berkeley. *Urban Design International*, 4: 35–37.

Bruce, V., Green, P.R. and Georgeson, M.A. (1996) *Visual perception, physiology, psychology and ecology*, East Sussex, Psychology Press.

Bureau of Land Management (1980) Visual simulation techniques, US Dept. of the Interior, Washington, D.C.: U.S. Government Printing Office (Stock no. 024-001-00116-6).

Buttenfield, B.P. and Ganter, J.H. (1990) Visualization and GIS: what should we see? What might we miss? *Proceedings of the 4th International Symposium on Spatial Data Handling*, 307–316.

Comment, B. (2000) Das Panorama. Nicolai, Berlin.

Cruz-Neira, C. (2003) Computational humanities: the new challenge for VR. *IEEE Computer Graphics and Applications*, May/June 2003, 10–13.

Danahy, J.W. (2001) Technology for dynamic viewing and peripheral vision in landscape visualization. *Landscape and Urban Planning*, 54: 125–138.

Dangermond, J., Hardison, L. and Smith, L.K. (1981) Some trends in the evolution of geographic information system technology. Harvard Library of Computer Graphics, 1981 Mapping Collection, 27–34.

Daniel, T.C. and Meitner, M.J. (2000) Representational validity of landscape visualizations: the effects of graphic realism on perceived scenic beauty of forest vistas. *Journal of Environmental Psychology*, 21: 61–72.

Darken, R.P., Cockayne, W.R. and Carmein, D. (1997) The omni-directional treadmill: a locomotion device for virtual worlds. *User Interface Software and Technology*, Banff, Canada, October 14–17, ACM, 213–221.

Delano Smith, C. (1987) Cartography in the prehistoric period in the Old world: Europe, the Middle East and North Africa in *History of Cartography, Volume I, Cartography in Prehistoric, Ancient and Medieval Europe and the Mediterranean*. J.B. Harley and D. Woodward (eds), Chicago/London: University of Chicago press, 54–101.

Deussen, O. (2003) A framework for geometry generation and rendering of plants with applications in landscape architecture, *Landscape and Urban Planning*, 64: 105–113.

Deussen, O., Hanrahan, P., Lintermann, B., Mech, R., Pharr, M. and Prusinkiewicz, P. (1998) *Realistic Modeling and Rendering of Plant Ecosystems*. Computer Graphics Proceedings, Annual Conference Series, ACM SIGGRAPH, 275–286.

DiBiase, D. (1990) Visualization in the Earth Sciences. *Earth and Mineral Sciences*, 59: 13–18.

DiBiase, D., MacEachren, A.M., Krygier, J.B. and Reeves, C. (1992) Animation and the role of map design in scientific visualization. *Cartography and Geographic Information Systems*, 19, No. 4: 201–214.

Doellner, J. and Hinrichs, K. (2002) A Generic Rendering System. *IEEE Transactions on Visualization and Computer Graphics*, 8(2), 99–118.

Dougenik, J.A. (1979) ASPEX Program Summary Description. Mapping Software and Cartographic Data Bases. Harvard Library of Computer Graphics, Mapping Collection, Volume Two, 49–56.

Dürer, A. (1525) Underweysung der messung, mit dem zirckel uñ richtscheyt in *Linien ebnen unnd gantzen corporen/durch Albrecht Dürer zusamen getzoge und zu nutz alle kunstlieb habenden mit zu gehörigen figuren in truck gebracht*. Nüremberg.

Ervin, S.M. (2001) Digital landscape modeling and visualization: a research agenda. *Landscape and Urban Planning*, 54: 49–62.

Faintich, M. (1980) Digital Image Technology. Mapping, charting and geodesy impact. Harvard Library of Computer Graphics, Mapping Collection, 29–40.

Fairbairn, D. and Taylor, G. (2002) Data collection issues in virtual reality for urban geographical representation and modelling in *Virtual Reality in Geography*. P. Fisher and D. Unwin (eds), London: Taylor and Francis, 220–238.

Farenc, N., Musse, S.R., Schweiss, E., Kallmann, M., Aune, O., Boulic, R. and Thalmann, D. (2000) A paradigm for controlling virtual humans in urban environment simulations. *Applied Artificial Intelligence*, 14: 69–91.

Faust, N.L., Jordan, L.E. and Furman, M.D. (1981) Development and implementation of a low-cost microcomputer system for LANDSAT analysis and geographic database applications. Harvard Library of Computer Graphics, 1981 Mapping Collection, 107–110.

Fournier, A. and Reeves, W.T. (1986) A Simple Model of Ocean Waves. *SIGGRAPH '86* 20, N4: 75–84.

French, I.A., Anderson, D.H. and Catchpole, E.A. (1990) Graphical simulation of bushfire spread. *Mathematical Computer Modelling*, 12: 67–71.

Fröhlich, T. (2000) *The virtual oceanarium*. Communications of the ACM. Vol. 43, No. 7, 94–101.

Gaggioli, A. (2001) Using virtual reality in experimental psychology in *Towards Cyberpsychology: mind, cognition and society in the Internet age*. G. Riva and C. Galimberti (eds), Amsterdam: IOS Press, 157–174.

Geyer, B. (1994) *Scheinwelten. Die Geschichte der Perspektive*, Leipzig: Seemann.

Hamilton, M.P. and Flaxman, M. (1992) Scientific data visualization and biological diversity: new tools for spatializing multimedia observations of species and ecosystems. *Landscape and Urban Planning*, 21, N 4: 285–287.

Hanson, M.G. and Lynch, D. (1979) Automated geocoding and display in map form of data survey. Mapping Software and Cartographic Data Bases. Harvard Library of Computer Graphics, Mapping Collection, Volume Two, 107–110.

Hedley, N.R., Billinghurst, M., Postner, L., May, R. and Kato, H. (2002) Explorations in the use of augmented reality for geographic visualization. *Presence: Teleoperators and Virtual Environments*, 11: 119–133.

Hehl-Lange, S. (2001) Structural elements in the visual landscape and their ecological functions. *Landscape and Urban Planning*, 54: 105–114.

Hoinkes, R. and Lange, E. (1995) 3D for free. Toolkit expands visual dimensions in GIS. *GIS World*, 8(7): 54–56.

House, D., Schmidt, G., Arvin, S. and Kitagawa-DeLeon, M. (1998) Visualizing a real forest. *IEEE Computer Graphics and Applications*, 18: 12–15.

Hull, R.B.I. and McCarthy, M.M. (1988) Change in the landscape. *Landscape and Urban Planning*, 15: 265–278.

Iwata, H. and Fujii, T. (1996) Virtual perambulator: a novel interface device for locomotion in virtual environment. *Proceedings of IEEE Virtual Reality Annual International Symposium*, Santa Clara, CA.

Kaneda, K., Okamoto, T., Nakamae, E. and Nishita, T. (1991) Photorealistic image synthesis for outdoor scenery under various atmospheric conditions. *The Visual Computer*, 7: 247–258.

Kwartler, M. and Bernard, R.N. (2001) CommunityViz: an integrated planning support system in *Planning Support Systems: integrating geographic information systems and visualization tools*. R.K. Brail and R.E. Klosterman (eds), Redlands, CA: ESRI Press, 285–308.

Lang, L. (1989) GIS Goes 3D. *Computer Graphics World*, 12: 38–46.

Lange, E. (1990) Vista management in Acadia National Park. *Landscape and Urban Planning*, 19: 353–376.

Lange, E. (1999) Realität und computergestützte visuelle Simulation. Eine empirische Untersuchung über den Realitätsgrad virtueller Landschaften am Beispiel des Talraums Brunnen/Schwyz. ORL-Berichte Nr. 106, VDF, Zürich, 176 pp.

Lange, E. (2001a) The limits of realism: perceptions of virtual landscapes. *Landscape and Urban Planning*, 54: 163–182.

Lange, E. (2001b) Prospektive 3D-Visualisierungen der Landschaftsentwicklung als Grundlage für einen haushälterischen Umgang mit der Ressource Landschaft. Natur und Landschaft, 76: 513–519.

Lange, E., Thoma, M. and Weber, G. (2001) Ländliche Strukturverbesserung und Landschaftsbild: Eine empirische Studie zur Beurteilung der Wirkung der Melioration mit Hilfe der 3D-Visualisierung. Paar, P. and Stachow, U. (eds), Visuelle Ressourcen – übersehene ästhetische Komponenten in der Landschaftsforschung und -entwicklung. *ZALF-Berichte*, 44: 88–95.

Lovett, A.A., Kennaway, J., Sünnenberg, G., Cobb, D., Dolman, P., O'Riordan, T. and Arnold, D. (2002) Visualizing sustainable agricultural landscapes in Virtual Reality in Geography. Fisher, P. and Unwin, D. (eds), London: Taylor & Francis: 102–130.

MacEachren, A.M. and Ganter, J.H. (1990) A pattern identification approach to cartographic visualization. *Cartographica*, 27: 64–81.

Magnenat-Thalmann, N., De Angelis, M., Hong, T. and Thalmann, D. (1989) Design, Transformation and Animation of Human Faces, *The Visual Computer*, 5: 32–39.

Magnenat-Thalmann, N. and Thalmann, D. (eds) (2001) *Deformable avatars*, Boston: Kluwer Academic Publishers, 247 pp.

Magnenat-Thalmann, N., Primeau, E. and Thalmann, D. (1987) Abstract muscle action procedures for human face animation, *The Visual Computer*, 3: 290–297.

Manandhar, D. and Shibasaki, R. (2001) Vehicle-borne laser mapping system (VLMS) for 3D urban GIS database. *Computers in Urban Planning and Urban Management*, Honolulu, HI, University of Hawaii.

McCormick, B.H., DeFanti, T.A. and Brown, M.D. (1987) Visualization in scientific computing. *Computer Graphics*, 21.

Ministère de la Culture et de la Communication (2002) *The cave of Chauvet-Pont-d'Arc*. http://www.culture.fr/culture/arcnat/chauvet/en/index.html

Molnar, D.J. (1986) SCEEN: An interactive computer graphics design system for real-time environmental simulation. *Landscape Journal*, 5: 128–134.

Moore, R. (1990) Landscapes on Pluto: improving computer-aided visualisation. *The Cartographic Journal*, 27: 132–136.

Myklestad, E. and Wagar, J.A. (1977) PREVIEW: computer assistance for visual management of forested landscapes. *Landscape Planning*, 4: 313–331.

Nadeau, D.R. (1999) Building virtual worlds with VRML. *IEEE Computer Graphics and Applications*, March/April, 18–29.

Nakamae, E., Qin, X. and Tadamura, K. (2001) Rendering of landscapes for environmental assessment. *Landscape and Urban Planning*, 54: 19–32.

Nebiker, S. (2003) Support for visualisation and animation in a scalable 3D GIS environment – motivation, concepts and implementation. International Archives of the Photogrammetry, Remote Sensing and Spatial Information Sciences, Vol. XXXIV-5/W10.

Oh, K. (1994) A perceptual evaluation of computer-based landscape simulations. *Landscape and Urban Planning*, 28: 201–216.

Orland, B. (1988) Video-imaging: A powerful tool for visualization and analysis. *Landscape Architecture*, 78–88.

Oettermann, S. (1980) Das Panorama: Die Geschichte eines Massenmediums. Syndikat, Frankfurt am Main.

Pair, J., Neumann, U., Piepol, D. and Swartout, B. (2003) FlatWorld: Combining Hollywood set-design techniques with VR. *IEEE Computer Graphics and Applications*, January/February, 12–15.

Peltz, D.L. and Kleinman, N.H. (1990) Visualization on the Macintosh. *Computer Graphics Review*, 20–39.

Perrin, L., Beauvais, N. and Puppo, M. (2001). Procedural landscape modeling with geographic information: the IMAGIS approach. *Landscape and Urban Planning*, 54: 33–48.

Pietsch, S.M. (2000) Computer visualisation in the design control of urban environments: a literature review. *Environment and Planning B: Planning and Design*, 27: 521–536.

Prusinkiewicz, P. and Lindenmayer, A. (1996) *The Algorithmic Beauty of Plants*, New York: Springer.

Reeves, W.T. and Blau, R. (1985) Approximate and probabilistic algorithms for shading and rendering structured particle systems. SIGGRAPH '85: 313–322.

Reffye de, P., Edelin, C., Françon, J., Jaeger, M. and Puech, C. (1988) Plant models faithful to botanical structure and development. SIGGRAPH 88, Computer Graphics, 22(4): 151–158.

Repton, H. (1803) *Observations on the Theory and Practice of Landscape Gardening: Including some Remarks on Grecian and Gothic Architecture*. London: Taylor; Oxford: Phaidon, 1980 (facs.).

Rhyne, T., Bolstad, M. and Rheingans, P. (1993) Visualizing environmental data at the EPA. *IEEE Computer Graphics and Applications*, 13(2): 34–38.

Ribarsky, W., Wasilewski, T. and Faust, N. (2002) From urban terrain models to visible cities. *IEEE Computer Graphics and Applications*, 22: 10–15.

Ribarsky, W., Bolter, J., Op den Bosch, A. and van Teylingen, R. (1994) Visualization and analysis using virtual reality. *IEEE Computer Graphics and Applications*, 10–12.

Rickenbacher, M. (2001a) Zahlenberge. Das Panorama im digitalen Zeitalter. In: Augenreisen. Das Panorama in der Schweiz. Hrsg.: Schweizerisches Alpines Museum, Schweizer Alpen-Club SAC, 150–173.

Rickenbacher, M. (2001b) Weltrekord! Das DIGIRAMA der Schweizer Alpen. *Vermessung, Photogrammetrie, Kulturtechnik*, 8: 544–549.

Rolland, J., Davis, L., Ha, Y., Meyer, C., Shaoulov, V., Akcay, A., Zheng, H., Banks, R. and Del Vento, B. (2002) 3D visualization and imaging in distributed collaborative environments. *IEEE Computer Graphics and Applications*, 22(1): 11–13.

Sutherland, I.E. (1968) A head-mounted three-dimensional display. *AFIPS Conference Proceedings*, 757–764.

Thomas, B.H. and Piekarski, W. (2003) Outdoor virtual reality. *1st International Symposium on Information and Communication Technologies*, Dublin, Trinity College Dublin, 226–231.

van Veen, H.A., Distler, H.K., Braun, S.J. and Bulthoff, H.H. (1998) Navigating through a virtual city: Using virtual reality technology to study human action and perception. *Future Generation Computer Systems*, 14: 231–242.

Verbree, E., van Maren, G., Germs, R., Jansen, F. and Kraak, M.-J. (1999) Interaction in virtual world views – linking 3D GIS with VR. *International Journal of Geographical Information Science*, 13: 385–396.

Vining, J. and Orland, B. (1989) The video advantage: A comparison of two environmental representation techniques. *Journal of Environmental Management*, 29: 275–283.

Walter, M., Fournier, A. and Menevaux, D. (2001) Integrating shape and pattern in mammalian models. Computer Graphics Proceedings, Annual Conference Series, ACM SIGGRAPH, 317–326.

White, W.B. (1992) Future for visualization through the integrated forest resource management system (INFORMS), *Landscape and Urban Planning*, 21: 277–279.

You, S., Neumann, U. and Azuma, R. (1999) Orientation tracking for outdoor augmented reality registration. *IEEE Computer Graphics and Applications*, 19: 36–42.

ZALF (2002) Machbarkeitsstudie für ein Visualisierungstool. http://www.zalf.de/dbu-vis/

Zimmerman, T.G. and Lanier, J. (1987) Hand gesture interface device. *CHI + GI Conference Proceedings*, 189–192.

Zoll, M. and Rosenberg, D. (1990) Of pixels and polymers. *CGW* 13: 103–106.

Zube, E.H., Simcox, D.E. and Law, C.S. (1987) Perceptual landscape simulations: History and prospect. *Landscape Journal*, 6: 62–80.

Chapter 2

Andrienko, G.L. and Andrienko, N.V. (1999) Interactive maps for visual spatial exploration. *International Journal of Geographical Information Science*, 13: 355–374.

Batara, A., Dave, B. and Bishop, I.D. (2001) Translation between multiple representations of spatial data. AURISA 2001, Melbourne, Australia, Australasian Urban and Regional Information Systems Association.

Bishop, I.D. (1994) The role of visual realism in communicating and understanding spatial change and process. *Visualization in Geographic Information Systems*. H. M. Hearnshaw and D. J. Unwin (eds) Chichester: John Wiley, 60–64.

Gaarder, J. (1996) *Sophie's World: a novel about the history of philosophy*, London: Phoenix (translated by Paulette Møller).

Germs, H.M.L., Maren, G.V., Verbree, E. and Jansen, F.W. (1999) A multi-view VR interface for 3D GIS. *Computers and Graphics*, 23: 497–506.

Graf, K.C., Suter, M., Hagger, J., Meier, E., Meuret, P. and Nuesch, D. (1994) Perspective terrain visualization – a fusion of remote sensing, GIS and computer graphics. *Computers and Graphics*, 18: 795–802.

References

Haughton, G. (1999) Information and participation within environmental management. *Environment and Urbanization,* 11: 51–62.

Hehl-Lange, S. (2001) Structural elements in the visual landscape and their ecological functions. *Landscape and Urban Planning,* 54: 105–114.

Hehl-Lange, S. and Lange, E. (1999) Planen mit virtuellen Braunkohlelandschaften. *Naturschutz und Landschaftsplanung,* 31(10): 301–307.

Heim, M. (1998) *Virtual Realism,* New York: Oxford.

Jacobson, L. (1993) *Garage Virtual Reality,* Indianaplois, IN: Sams Publishing.

Krause, C.L. (2001) Our visual landscape: managing the landscape under special consideration of visual aspects. *Landscape and Urban Planning,* 54: 239–254.

Lange, E. (1994) Integration of computerized visual simulation and visual assessment in environmental planning. *Landscape and Urban Planning,* 30: 99–112.

Langendorf, R. (2001). Computer-aided visualization: possibilities for urban design, planning and management. *Planning Support Systems: integrating geographic information systems, models and visualization tools.* R.K. Brail and R.E. Klosterman (eds), Redlands, CA: ESRI Press, 309–359.

MacEachren, A.M., Bishop, I.D., Dykes, J., Dorling, D. and Gatrell, A. (1994) Introduction to advances in visualizing spatial data. *Visualization in geographic Information systems.* H. M. Hearnshaw and D. J. Unwin (eds) Chichester: John Wiley, 51–59.

MacEachren, A. M., Edsall, R., Haug, D., Baxter, R., Otto, G., Masters, R., Fuhrmann S. and Qian, L. (1999) Virtual environments for geographic visualisation: potential and challenges. *ACM Workshop on New Paradigms for Information Visualization and manipulation,* Kansas City, KS (35–40).

McCormick, B.H., DeFanti, T.A. and Brown, M.D. (1987) Visualization in scientific computing. *Computer Graphics,* 21.

Perrin, L., Beauvais, N. and Puppo, M. (2001) Procedural landscape modeling with geographic information: the IMAGIS approach. *Landscape and Urban Planning,* 54: 33–48.

Robertson, P.K. (1991) A methodology for choosing data representations. *IEEE Computer Graphics and Applications,* May: 56–67.

Sherman, B. and Judkins, P. (1992) *Glimpses of Heaven, Visions of Hell: virtual reality and its implications.* UK: Hodder & Stoughton.

Sherman, W.R. and Craig, A.B. (2003) *Understanding Virtual Reality.* Amsterdam: Morgan Kaufmann.

Slocum, T.A., Blok, C., Jiang, B., Koussoulakou, A., Montello, D.A., Fuhrmann, S. and Hedley, N.R. (2001) Cognitive and usability issues in geovisualization. *Cartography and Geographic Information Science,* 28: 61–75.

Tufte, E.R. (1990) *Envisioning Information.* Connecticut: Graphics Press, Cheshire.

Verbree, E., van Maren, G., Germs, R., Jansen, F. and Kraak, M.-J. (1999) Interaction in virtual world views – linking 3D GIS with VR. *International Journal of Geographical Information Science,* 13: 385–396.

Winn, W., Windschitl, M., Fruland, R. and Lee, Y.-l. (2002) Features of virtual environments that contribute to learners' understanding of Earth Science. National Association for Research in Science Teaching, New Orleans, April 2002.

Zube, E.H., Simcox, D.E. and Law, C.S. (1987) Perceptual landscape simulations: History and prospect. *Landscape Journal,* 6: 62–80.

Chapter 3

Ogleby, C.L. and Kenderdine, S. (2001) Ancient Olympia as a three-dimensional museum experience. *Cultural Heritage and Technologies in the Third Millennium*, Proceedings of the International Cultural Heritage Informatics Meeting, Politechnico di Milano, Italy, 3–7 September, 333–340.

Chapter 4

Introduction

Moore, G.E. (1965) Cramming more components onto integrated circuits. *Electronics*, 38.

Efficient modelling and rendering of landscapes

Aono, M. and Kunii, T.L. (1984) Botanical Tree Image Generation. *IEEE Computer Graphics and Applications*, 4: 10–34.

Bloomenthal, J. (1985) Modeling the Mighty Maple. In B.A. Barsky, Hrsg., Computer Graphics (Proceedings SIGGRAPH '85), 19: 305–311.

Deussen, O., Colditz, C., Stamminger, M. and Drettakis, G. (2002) Interactive visualization of complex plant ecosystems in IEEE Visualization 2002, S. 219–226.

Deussen, O., Hanrahan, P., Pharr, M., Lintermann, B., Mech, R. and Prusinkiewicz, P. (1998) Realistic modeling and rendering of plant ecosystems in SIGGRAPH '98 Conference Proceedings, 275–86.

Holton, M. (1994) Strands, gravity and botanical tree imagery. *Computer Graphics Forum*, 13: 57–67.

Lintermann, B. and Deussen, O. (1999) Interactive modelling of plants. *IEEE Computer Graphics and Applications*, 19: 56–65.

Mech, R. and Prusinkiewicz, P. (1996) Visual models of plants interacting with their environment in Proceedings of SIGGRAPH 1996, 397–410.

Oppenheimer, P. (1986) Real time design and animation of fractal plants and trees. In Computer Graphics (SIGGRAPH '86 Proceedings), 20: 55–64.

Prusinkiewicz, P. and Lindenmayer, A. (1990) The algorithmic beauty of plants. New York: Springer-Verlag.

Prusinkiewicz, P., Mündermann, L., Karwowski, R. and Lane, B. (2001) The use of positional information in the modelling of plants in Proceedings of SIGGRAPH 2001, 289–300.

Thatcher, U. (2002) Rendering massive terrains using chunked level of detail, SIGGRAPH 2002 course material.

Using games software for interactive landscape visualization

Herwig, A. and Paar, P. (2002) Game engines: Tools for landscape visualization and planning? in Erich Buhmann, Ursula Nothelfer and Matthias Pietsch (eds) *Trends in GIS and virtualization in environmental planning and design*, Proceedings at Anhalt University of Applied Sciences, Wichmann, Heidelberg, 162–171.

Kretzler, E. (2003) Improving landscape architecture design using real time engines, in E. Buhmann and S. Ervin (eds) *Trends in landscape modeling*, Proceedings at Anhalt University of Applied Sciences, Wichmann, Heidelberg, 95–101.

Orland, B., Ogleby, C., Bishop, I., Campbell, H. and Yates, P. (1997): Multi-media approaches to visualization of ecosystem dynamics, in Proceedings, ASPRS/

ACSM/ RT '97, Seattle, American Society for Photogrammetry and Remote Sensing, Washington, DC, vol. 4, 224–36.

Presentation style and technology

Anderson, L.M., Mulligan, B.E., Goodman, L.S. and Regen, H.Z. (1983) Effects of sounds on preferences for outdoor settings. *Environment and Behavior*, 15: 539–566.

Appleton, K. and Lovett, A. (2003) GIS-based visualisation of rural landscapes: defining 'sufficient' realism for environmental decision-making. *Landscape and Urban Planning*, 65: 117–131.

Bishop, I.D. (2001) Predicting movement choices in virtual environments. *Landscape and Urban Planning*, 56: 97–106.

Bishop, I.D. (2002) Determination of thresholds of visual impact: the case of wind turbines. *Environment and Planning B: Planning and Design*, 29: 707–718.

Bishop, I.D. and Hulse, D.W. (1994) Predicting scenic beauty using mapping data and geographic information systems. *Landscape and Urban Planning*, 30: 59–70.

Bishop, I.D. and Rohrmann, B. (2003) Subjective responses to simulated and real environments: a comparison. *Landscape and Urban Planning*, 65: 261–277.

Bishop, I.D., Fasken, G., Ford, R., Hickey, J., Loiterton, D. and Williams, K. (2003) Visual simulation of forest regrowth under different harvest options. *Trends in landscape modelling*: Proceedings at Anhalt University of Applied Sciences, Dessau, Germany, May 15–16, Herbert Wichmann Verlag, 46–55.

Bishop, I.D., Wherrett, J.R. and Miller, D. (2001a) Assessment of path choices on a country walk using a virtual environment. *Landscape and Urban Planning*, 52: 227–239.

Bishop, I.D., Ye, W.-S. and Karadaglis, C. (2001b) Experiential approaches to perception response in virtual worlds. *Landscape and Urban Planning*, 54: 115–124.

Bowman, D., Davis, E., Badre, A. and Hodges, L. (1999) Maintaining spatial orientation during travel in an immersive virtual environment. *Presence: Teleoperators and Virtual Environments*, 8: 618–631.

Carles, J.L., Barrio, I.L. and Lucio, J.V.d. (1999) Sound influence on landscape values. *Landscape and Urban Planning*, 43: 191–200.

Champion, E., Dave, B. and Bishop, I.D. (2003) Evaluation of interaction vs. engagement for virtual heritage applications. CAADFUTURES 2003, Taiwan: Kluwer Academic Publishers

Cruz-Neira, C., Sandin, D.J., Fanti, T.A.D. and Hart, J.C. (1992) The Cave: Audio visual experience automatic virtual environment. *Communications of the ACM*, 35: 64–72.

Daniel, T.C. and Vining, J. (1983) Methodological issues in the assessment of landscape quality. *Behavior and the Natural Envionment*, I. Altman and J. F. Wohlwill. New York: Plenum: 39–84.

Darken, R.P. and Sibert, J.L. (1996) Navigating in large virtual worlds. *The International Journal of Human–Computer Interaction*, 8: 49–72.

Dodge, M. (1999) Exploration in AlphaWorld: the geography of 3-D virtual worlds on the Internet. iVirtual Reality in Geography, RGS-IBG Conference, Leicester (http://www.casa.ucl.ac.uk/martin/ibg99.pdf).

Draper, J.V., Kaber, D.B. and Usher, J.M. (1998) Telepresence. *Human Factors*, 40: 354–375.

Esposito, C. and Orland, B. (1984) An investigation into the role of sound in landscape perception. Council of Educators in Landscape Architecture, Guelph, Ontario, Canada (CELA).

Hetherington, J., Daniel, T.C. and Brown, T.C. (1993) Is motion more important than it sounds?: The medium of presentation in environment perception research. *Journal of Environmental Psychology*, 13: 283–291.

Lange, E. (2001) The limits of realism: perceptions of virtual landscapes. *Landscape and Urban Planning*, 54: 163–182.

Lange, E., Thoma, M. and Weber, G. (2001) Ländliche Strukturverbesserung und Landschaftsbild: Eine empirische Studie zur Beurteilung der Wirkung der Melioration mit Hilfe der 3D-Visualisierung, in P. Paar and U. Stachow (eds) *Visuelle Ressourcen – übersehene ästhetische Komponenten in der Landschaftsforschung und -entwicklung*. ZALF-Berichte, 44: 88–95.

Langendorf, R. (2001) Computer-aided visualization: possibilities for urban design, planning and management. *Planning Support Systems: integrating geographic information systems, models and visualization tools*. R. K. Brail and R. E. Klosterman, Redlands, CA: ESRI Press, 309–359.

MacEachren, A.M., Edsall, R., Haug, D., Baxter, R., Otto, G., Masters, R., Fuhrmann, S. and Qian, L. (1999) Virtual environments for geographic visualisation: potential and challenges. *Proceedings of New Paradigms in Information Visualization and Manipulation*, Kansas City, MO, 35–40.

Meitner, M.J. (2004) Scenic beauty of river views in the Grand Canyon: relating perceptual judgments to locations. *Landscape and Urban Planning*, 68: 3–13.

Nadeau, D.R. (1999) Building virtual worlds with VRML. *IEEE Computer Graphics and Applications*, March/April 1999: 18–29.

Palmer, J.F. and Hoffman, R.E. (2001) Rating reliability and representation validity in scenic landscape assessments. *Landscape and Urban Planning*, 54: 149–162.

Petschek, P. (2003) Planning for public spaces – the application of new media and 3D visualizations: the Zürich-Leutschenbach case study. *Trends in landscape modeling*, Dessau: Herbert Wichmann Verlag, 158–162.

Slater, M. and Steed, A. (2000) A virtual presence counter. *Presence, Teleoperators and Virtual Environments*, 9: 413–434.

van Veen, H.A., Distler, H.K., Braun, S.J. and Bulthoff, H.H. (1998) Navigating through a virtual city: Using virtual reality technology to study human action and perception. *Future Generation Computer Systems*, 14: 231–242.

Wherrett, J.R. (1999) Issues in using the Internet as a medium for landscape preference research. *Landscape and Urban Planning*, 45: 209–217.

Zube, E.H., Sell, J.L. and Taylor, J.G. (1982) Landscape Perception: Research, Application and Theory. *Landscape Planning*, 9: 1–33.

Chapter 5

Appleyard, D. (1977) Understanding professional media: issues, theory, and a research agenda, in *Human Behavior and Environment*, I. Altman and J.F. Wohlwill (eds), New York: Plenum Press, 43–88.

Bergen, S.D., Ulbricht, C.A., Fridley, J.L. and Ganter, M.A. (1995) The validity of computer generated graphic images of forest landscapes. *Journal of Environmental Psychology*, 15: 135–146.

Bishop, I.D. (2001) Predicting movement choices in virtual environments. *Landscape and Urban Planning*, 56: 97–106.

Bishop, I.D. and Leahy, P.N.A. (1989) Assessing the visual impact of development proposals: the validity of computer simulations. *Landscape Journal*, 8: 92–100.

Cavens, D. (2002) A design-based decision support system for forest design. Unpublished MSc. thesis, University of British Columbia, Vancouver, BC.

References

Craik, K.H., Appleyard, D. and McKechnie, G.E. (1980) Impressions of a place: effects of media and familiarity among environmental professionals. Research Technical Report, Berkeley, CA: Institute of Personality Assessment and Research.

Danahy, J. (2001a) Technology for dynamic viewing and peripheral vision in landscape visualization. *Landscape and Urban Planning*, 54: 125–137

Danahy, J. (2001b) Considerations for digital visualization of landscape. Chapter 15, 225–45, in Sheppard, S.R.J. and Harshaw, H. (eds), *Forests and Landscapes: Linking Sustainability, Ecology, and Aesthetics*. International Union of Forest Research Organizations (IUFRO) Research Series. Wallingford, UK: CABI International.

Daniel, T.C., Meitner, M. and Orland, B. (1997) Public perceptions and expert assessments: the effects of false color in forest visualizations. Project final report. USDA Forest Service, Forest Health Protection, Forest Health Technology Enterprise Team.

Daniel, T.C. and Meitner, M.J. (2001) Representational validity of landscape visualizations: the effect of graphical realism on perceived scenic beauty of forest vistas. *Journal of Environmental Psychology*, 21: 61–72.

Lange, E. (2001) The limits of realism: perceptions of virtual landscapes. *Landscape and Urban Planning*, 54: 163–182.

Lewis, J.L. (2000) Ancient values, new technology: Emerging methods for integrating cultural values in forest management. Unpublished MSc. thesis, Faculty of Forestry, UBC, Vancouver, BC.

Luymes, D.T. (2001) The rhetoric of visual simulation in forest design: Some research directions. Chapter 13, 191–204, in Sheppard, S.R.J. and Harshaw, H. (eds), *Forests and Landscapes: Linking Sustainability, Ecology, and Aesthetics*. International Union of Forest Research Organizations (IUFRO) Research Series. Wallingford, UK: CABI International.

McQuillan, A.G. (1998) Honesty and foresight in computer visualizations. *Journal of Forestry*, 96: 15–16.

Meitner, M. and Daniel, T. (1997) The effects of animation and interactivity on the validity of human responses to forest data visualizations, in Orland, B. (ed.) *Proceedings of Data Visualization '97*, St Louis, MO.

Oh, K. (1994) A perceptual evaluation of computer-based landscape simulations. *Landscape and Urban Planning*, 28: 201–216.

Orland B. (1992) Data visualization techniques in environmental management: A consolidated research agenda. *Landscape and Urban Planning*, 21: 241–244.

Orland, B. (1997) Forest virtual modelling for planners and managers. Proceedings ASPRS/RTI'97, Seattle, April 1997.

Orland, B. and Uusitalo, J. (2001) Immersion in a virtual forest: some implications. Chapter 14, 205–24, in Sheppard, S.R.J. and Harshaw, H. (eds), *Forests and Landscapes: Linking Sustainability, Ecology, and Aesthetics*. International Union of Forest Research Organizations (IUFRO) Research Series. Wallingford, UK: CABI International.

Orland, B., Budthimedhee, K. and Uusitalo, J. (2001) Considering virtual worlds as representations of landscape realities and as tools for landscape planning. *Landscape and Urban Planning*, 54: 139–148.

Palmer, J., Pitt, D.G. and Freimund, W.A. (1995) The virtual landscape: converging trends from artificial intelligence, geographical information systems, and digital visualization technologies in visual landscape assessment. *Proceedings of the 4th International Outdoor Recreation and Tourism Trends Symposium*, May 14–17, St Paul, MN, University of Minnesota.

Palmer, J.F. and Hoffman, R.E. (2001) Rating reliability and representation validity in scenic landscape assessments. *Landscape and Urban Planning*, 54: 149–161.

Sheppard, S.R.J. (1982) Landscape portrayals: their use, accuracy, and validity in simulating proposed landscape change. PhD dissertation, University of California, Berkeley, CA.

Sheppard, S.R.J. (1986) Simulating changes in the landscape, in *Foundations for Visual Project Analysis*, Smardon, R.C., Palmer, J. and Felleman, J. (eds) Chapter 11, 187–99. New York: John Wiley and Sons.

Sheppard, S.R.J. (1989) *Visual simulation: A user's guide for architects, engineers, and planners*. New York: Van Nostrand Reinhold.

Sheppard, S.R.J. (2000) Visualization as a decision-support tool for managing forest ecosystems. *The COMPILER*, 16: 25–40.

Sheppard, S.R.J. (2001) Guidance for crystal ball gazers: developing a code of ethics for landscape visualization. *Landscape and Urban Planning*, 54: 183–199.

Taylor, M. and McDaniel, R. (1997) Benefits of design visualization tools, in *Proc. Establishing the Worth of Scenic Values: The Tahoe Workshop*, Strain, A. and Sheppard, S.R.J. (eds), Lake Tahoe, October, 1997.

Weller, S.C. and Romney, A.K. (1988) *Systematic Data Collection*. Qualitative Research Methods Series, Vol. 10. Newbury Park: Sage Publications, 96 pp.

Wherrett, J.R. (2001) Predicting preferences for scenic landscapes using computer simulations. Chapter 16, 247–60, in Sheppard, S.R.J. and Harshaw, H. (eds), *Forests and landscapes: Linking sustainability, ecology, and aesthetics*. International Union of Forest Research Organizations (IUFRO) Research Series. Wallingford, UK: CABI International.

Chapter 6

Visualization in forest landscapes

Kojima, M. and Wagar, J.A. (1972) Computer generated drawings of groundform and vegetation. *Journal of Forestry*, 70(5): 282–285.

Magill, A.W. (1992) Managed and natural landscapes: what do people like? USDA Forest Service Research Paper PSW-RP-213. Pacific Southwest Forest and Range Exp. Stn., Berkeley, CA. 28 pp.

McGaughey, R. (1997) Techniques for visualizing the appearance of timber harvest operations, *Proceedings of Council on Forest Engineering: Forest Operations for Sustainable Forests and Healthy Economies*, Rapid City, South Dakota, July 1997.

Myklestad, E. and Wagar, J.A. (1977) Preview: computer assistance for visual management of forested landscapes. *Landscape Planning*, 4: 313–331.

Picard, P. and Sheppard, S.R.J. (2001) Visual resource management in British Columbia: Part II. Partial cutting in the front-country: a win–win solution for short-term timber availability and aesthetics? *B.C. Journal of Ecosystem and Management*, Volume 1, No. 2.

Province of British Columbia, Ministry of Forests (BCMoF) (1996) *Clearcutting and visual quality: A public perception study*. Recreation Section, Range, Recreation and Forest Practices Branch, BCMoF, Victoria BC.

Province of British Columbia, Ministry of Forests (BCMoF) (1997) *Visual Landscape Inventory Procedures and Standards Manual*. Forest Practices Branch, BCMoF, Victoria BC.

United States Department of Agriculture Forest Service (1995) *Landscape aesthetics: A handbook for scenery management*. USDA Forest Service Agriculture Handbook No. 701. US Government Printing Office, Washington, DC.

'Calibrating' images to more accurately represent future landscape conditions in forestry

Daniel, T.C., Weideman, E. and Hines, D. (2003) Assessing public tradeoffs between fire hazard and scenic beauty in the wildland/urban interface, in (Jakes, P. Compiler) Homeowners, communities and wildfire: science findings from the National Fire Plan, USDA Forest Service, North Central Station, General Technical Report NC-231, 36–44.

Gilmore, D.W., Kastendick, D.N., Zasada, J.C. and Anderson, P.J. (2003) Alternative fuel reduction treatments in the Gunflint Corridor of the Superior National Forest: Second year results and sampling recommendations. Res. Note NC-381. St Paul, MN: US Department of Agriculture, Forest Service, North Central Research Station, 8 pp.

McGaughey, R. (2003) SVS – The Stand Visualization System, USDA Pacific Northwest Forest Experiment Station, Seattle. http://forsys.cfr.washington.edu/svs.html

Orland, B. (2003) SmartForest – Interactive Forest Visualization. Penn State University, Department of Landscape Architecture. http://www.imlab.psu.edu/smartforest/index.html

Orland, B., Vining, J. and Ebreo, A. (1992) The effect of street trees on perceived values of residential property, Environment and Behavior, 24(3): 298–325.

Teck, R., Moeur, M. and Eav, B. (1996) Forecasting ecosystems with the forest vegetation simulator. *Journal of Forestry*, 94: 7-10.

USDA Forest Service. (2000) Gunflint Corridor fuel reduction: final environmental impact statement. Gunflint Ranger District. Grand Marais, MN: Superior National Forest, Cook County.

Studying the acceptability of forest management practices using visual simulation of forest regrowth

Florence, R.G. (1996) *Ecology and silviculture of eucalypt forests*. CSIRO, Collingwood, Victoria.

Hickey, J.E., Neyland, M.G. and Bassett, O.D. (2001) Rationale and design for the Warra silvicultural systems trial in wet *Eucalyptus obliqua* forests in Tasmania. *Tasforests*, 13: 155–182.

Planning, communicating, designing and decision making for large scale landscapes

Al-Kodmany, K. (1999) Using visualization techniques for enhancing public participation in planning and design: Process, implementation, and evaluation. *Landscape and Urban Planning*, 45: 37–45.

Cavens, D. (2002) A design-based decision support system for forest design. Unpublished MSc. thesis, University of British Columbia, Vancouver, BC.

Cavens, D. and Sheppard, S.R.J. (2003) CALP Forester: An interactive forestry user interface for design-based decision support. Paper presented at IUFRO Decision Support for Multiple Purpose Forestry Conference, Vienna, Austria, 23–25 April 2003.

Daniel, T.C. and Meitner, M.J. (2001) Representational validity of landscape visualizations: the effect of graphical realism on perceived scenic beauty of forest vistas. *Journal of Environmental Psychology*, 21(1): 61–72.

Kimmins, J.P., Mailly, D. and Seely, B. (1999) Modeling forest ecosystem net primary production: the hybrid simulation approach used in FORECAST. *Ecological Modeling*, 122: 195–224.

Lewis, J.L. (2000) Ancient values, new technology: Emerging methods for integrating cultural values in forest management. Unpublished MSc. thesis, Faculty of Forestry, UBC, Vancouver, BC.

Meitner, M.J., Sheppard, S.R.J., Cavens, D., Gandy, R., Picard, P., Harshaw, H. and Harrison, D. (in press) The multiple roles of environmental/data visualization in evaluating alternative forest management strategies: Lessons learned, errors uncovered and questions raised. *Computers in Agriculture* (COMPAG) special issue, based on papers presented at IUFRO Decision Support for Multiple Purpose Forestry Conference Proceedings, Vienna, 2003.

McQuillan, A.G. (1998) Honesty and foresight in computer visualizations. *Journal of Forestry*, 96(6): 15–16.

Nelson, J. (2003) Forest-level models and challenges for their successful application. *Canadian Journal of Forest Research*, 33: 422–429.

Orland, B. and Uusitalo. J. (2001) Immersion in a virtual forest: some implications. Chapter 14 in Sheppard, S.R.J. and Harshaw, H. (eds), *Forests and landscapes: Linking sustainability, ecology and aesthetics*. International Union of Forest Research Organizations (IUFRO) Research Series. Wallingford, UK: CABI International, 205–224.

Sheppard, S.R.J. (2000) Visualization as a decision-support tool for managing forest ecosystems. *The COMPILER*, 16(1): 25–40.

Sheppard, S.R.J. (2001) Guidance for crystal ball gazers: developing a code of ethics for landscape visualization. *Landscape and Urban Planning*, 54(1–4): 183–199.

Sheppard, S.R.J. (2003) Knowing a socially sustainable forest when you see one: Implications for results-based forestry. *Forestry Chronicle*, 79(5): 865–875.

Sheppard, S.R.J. and Lewis, J.L. (2002) Democratising the SFM planning process: the potential of landscape visualization as a community involvement tool for First Nations, in Procs. 2002 Sustainable Forest Management Network Conference, *Advances in Forest Management: From Knowledge to Practice*. Veeman, T.S. et al. (eds) Sustainable Forest Management Network, Edmonton, Alberta, 304–309.

Sheppard, S.R.J. and Salter, J.D. (2004) The role of visualization in forest planning in *Encyclopedia of Forest Sciences*, Oxford, UK: Academic Press/Elsevier, 486–498.

Sheppard S.R.J. and Meitner, M.J. (2003) Using multi-criteria analysis and visualization for sustainable forest management planning with stakeholder groups, in Procs. Decision Support for Multiple Purpose Forestry IUFRO Conference, Vienna, Austria, March 2003.

Sheppard, S.R.J., Lewis, L.J. and Akai, C. (2002) Visualization for First Nations: Guidelines for use in aboriginal land planning and forest management. Technical Report, prepared for Sustainable Forest Management Network, Edmonton, Alberta. CALP, UBC, Vancouver, BC.

Wells, R.W. and Bunnell, F.L. (2001) Habitat analysis of Lemon Landscape Unit management scenarios. Technical Report, UBC Sustainability Project, prepared for Arrow Forest District Innovative Forest Practices Agreement: Nelson, BC.

The role of landscape simulators in forestry: a Finnish perspective

Karjalainen, E. and Tyrväinen, L. (2002) Visualization in landscape preference research: a Finnish perspective. *Landscape and Urban Planning*, 59: 13–28.

References

Karppinen, H., Hänninen, H. and Ripatti, P. (2002) Finnish forest owners 2000. The Finnish Forest Research Institute, Research Papers 852. 83 pp. (in Finnish).

McGaughey, R. (1997) Visualizing forest stand dynamics using the stand visualization system, in Proc. ACSM/ASPRS/RT Annual Convention, Seattle, Washington. American Society for Photogrammetry and Remote Sensing, Bethesda, Maryland, technical papers 4: 248–257.

Nousiainen, I., Tahvanainen, L. and Tyrväinen, L. (1998) Rural landscape in farm scale land-use planning. *Scandinavian Journal of Forest Research*, 13: 477–87.

Orland, B. (1988) Video-imaging: a powerful tool for visualization and analysis. *Landscape Architecture*, 76: 58–63.

Orland, B. (1994) SmartForest: 3-D interactive forest visualization and analysis, in Proc. Decision Support 2001, Resource Technology '94, Toronto. American Society for Photogrammetry and Remote Sensing, Bethesda, Maryland, 181–190.

Orland, B., Budthimedhee, K. and Uusitalo, J. (2000) Considering virtual worlds as representations of landscape realities and as tools for landscape planning. *Landscape and Urban Planning*, 54: 139–148.

Orland, B. and Uusitalo, J. (2000) Immersion in a virtual forest – some implications, 205–24, in Sheppard, S.R.J. and Harshaw, H.W. (eds) *Forests and landscapes: Linking ecology, sustainability and aesthetics*. IUFRO research Series, No. 6. CABI Publishing, Wallingford Oxon, UK. 304 pp.

Pukkala, T., Nuutinen, T. and Kangas, J. (1995) Integrating scenic and recreational amenities into numerical forest planning. *Landscape and Urban Planning*, 32: 185–195.

Rautalin, M., Uusitalo, J. and Pukkala, T. (2001) Estimation of tree stand characteristics through computer visualization. *Landscape and Urban Planning*, 53: 85–94.

Tahvanainen, L., Tyrväinen, L., Ihalainen, M., Vuorela, N. and Kolehmainen, O. (2001) Forest management and public perceptions – visual versus verbal information. *Landscape and Urban Planning*, 53: 53–70.

Tyrväinen, L. and Tahvanainen, L. (1999) Using computer graphics for assessing the scenic value of large-scale rural landscape. *Scandinavian Journal of Forest Research*, 14: 282–288.

Tyrväinen, L. and Tahvanainen, L. (2000) Landscape visualization in rural land-use planning, in Krishnapillay, B. *et al.* (eds) *Forests and society: The role of research*, XXI IUFRO World Congress, 7–12 August 2000, Kuala Lumpur, Malaysia. Volume 1, Sub-plenary sessions, 338–347.

Tyrväinen, L., Gustavsson, R., Konijnendijk, C. and Ode, Å. (in press) Visualization and landscape laboratories in developing planning, design and management of urban woodlands. Forest Policy and Economics. 15 pp.

Uusitalo, J., Orland, B. and Liu, K. (1997) Developing a forest manager interface to SmartForest. Proceedings GIS '97. Vancouver, February 17–20, 1997, 231–235.

Uusitalo, J. and Kivinen, V-P. (2000) Implementing SmartForest forest visualization tool on PC environment. Proceedings RT '98 Nordic. Rovaniemi, June 8–12, 1999. Finnish Forest Research Institute, research papers 791, 1–7.

Uusitalo, J. and Orland, B. (2001) Virtual forest management: Possibilities and challenges. *Journal of Forest Engineering*, 12: 57–66.

Chapter 7

Applications in the agricultural landscape – introduction

Angileri, V. and Toccolini, A. (1993) The assessment of visual quality as a tool for the conservation of rural landscape diversity. *Landscape and Urban Planning*, 24: 105–112.

Cooper, A. and Murray, R. (1992) A structured method of landscape assessment and countryside management. *Applied Geography*, 12: 319–338.

Gimblett, H.R. (1990) Environmental cognition: the prediction of preference in rural Indiana. *The Journal of Architectural and Planning Research*, 7, n3: 222–235.

Gomez-Limon, J. and de Lucio Fernandez, J.V. (1999) Changes in the use and landscape preferences in the agricultural-livestock landscapes of the central Iberian peninsular. *Landscape and Urban Planning*, 44: 165–175.

Hernandez, J., Garcia, L., Moran, J., Juan, A. and Ayuga, F. (2003) Estimating visual perception of rural landscapes: the influence of vegetation. The case of Elsa Valley (Spain), *Food, Agriculture and Environment*, 1: 139–141.

Hunziker, M. (1995) The spontaneous reafforestation in abandoned agricultural lands: perception and aesthetic assessment by locals and tourists. *Landscape and Urban Planning*, 31: 399–410.

Lynch, J.A. and Gimblett, H.R. (1992) Perceptual values in the cultural landscape: A computer model for assessing and mapping perceived mystery and rural environments. *Computers, Environment and Urban Systems*, 16: 453–471.

Orland, B. (1988) Aesthetic preference for rural landscapes: some resident and visitor difference, in *The Video Quality of The Environment: Theory, Research and Applications*, J. Nasar, Cambridge: 364–367.

Schauman, S. (1988) Countryside scenic assessment: tools and an application. *Landscape and Urban Planning*, 15: 227–239.

Tress, B. and Tress, G. (2003) Scenario visualisation for participatory landscape planning – a study from Denmark. *Landscape and Urban Planning*, 64: 161–178.

Vos, W. and Meekes, H. (1999) Trends in European cultural landscape development: perspectives for a sustainable future. *Landscape and Urban Planning*, 46: 3–14.

Zube, E.H. (1973) Rating everyday rural landscapes of the Northeastern U.S. *Landscape Architecture*, 63: 370–375.

Designing, visualizing and evaluating sustainable agricultural landscapes

Appleton, K., Lovett, A., Sünnenberg, G. and Dockerty, T. (2002) Rural landscape visualisation from GIS databases: a comparison of approaches, options and problems. *Computers, Environment and Urban Systems*, 26: 141–162.

Cobb, D., Dolman, P. and O'Riordan, T. (1999) Interpretations of sustainable agriculture in the UK. *Progress in Human Geography*, 23: 209–235.

Countryside Agency (2002) *The State of the Countryside 2002*. CA 109, Countryside Agency Publications, Wetherby.

Countryside Agency (2003) *The State of the Countryside 2020*. CA138, Countryside Agency Publications, Wetherby.

Department of the Environment, Transport and the Regions and Ministry of Agriculture, Fisheries and Food (2000) *Our countryside: the future – a fair deal for rural England*, Cm 4909, London: DETR.

Dolman, P., Lovett, A., O'Riordan, T. and Cobb, D. (2001) Designing whole landscapes. *Landscape Research*, 26: 305–335

Lovett, A., Kennaway, J., Sünnenberg, G., Cobb, D., Dolman, P., O'Riordan, T. and Arnold, D. (2002) Visualizing sustainable agricultural landscapes, in Fisher, P. and Unwin, D. (eds) *Virtual Reality in Geography*, London: Taylor & Francis, 102–130.

O'Riordan, T., Lovett, A., Dolman, P., Cobb, D. and Sünnenberg, G. (2000) Designing and implementing whole landscapes. *Ecos*, 21: 57–68.

Smith, S. (1997) A dimension of sight and sound. *Mapping Awareness*, October 1997, 18–21.

Smith, S. (1998) More than meets the eye. *Mapping Awareness*, July 1998, 24–27.

Helping rural communities envision their future

Chen, X., Bishop, I.D. and Abdul Hamid, A.R. (2002) Community exploration of changing landscape values: the role of the virtual environment, *Proceedings of Digital Image Computing – Techniques and Applications*, Melbourne, 273–278.

Stock, C. and Bishop, I.D. (2002) Immersive, interactive exploration of changing landscapes. *Proceedings IEMSs – Integrated Assessment and Decision Support*, Lugano, Switzerland, International Environmental Modelling and Software Society (iEMSs) 1, 30–35.

Lenné3D – walk-through visualization of planned landscapes

Lintermann, B. and Deussen, O. (1998) A modeling method and user interface for creating plants. *Computer Graphics Forum*, 17: 73–82.

Paar, P. (2003) Lenné3D – the making of a new landscape visualization system: From requirements analysis and feasibility survey towards prototyping, in Buhmann, E. and Ervin, S. (eds) *Trends in Landscape Modeling*, Wichmann, Heidelberg, 78–84.

Rekittke, J. (2002) Drag and drop – the compatibility of existing landscape theories and new virtual landscapes, in Buhmann, E., Nothelfer, U., and Pietsch M. (eds), *Trends in GIS and virtualization in environmental planning and design*. Wichmann, Heidelberg, 110–123.

Chapter 8

Applications in energy, industry and infrastructure: Introduction

Carson, R. (1962) *Silent Spring*, Boston: Houghton Mifflin.

Ehrlich, P.R. and Ehrlich, A.H. (1970) *Population, resources, environment: issues in human ecology*, San Francisco: W.H. Freeman and Co.

Sheppard, S.R.J. (1989) *Visual Simulation: A User's Guide for Architects, Engineers, and Planners*. New York, NY: Van Nostrand Reinhold.

Petrich, C.H. (1979) Aesthetic impact of a proposed power plant on an historic wilderness landscape. Our National Landscape – a conference on applied techniques for analysis and management of the visual resource, Incline Village, Nevada, UDSA Forest Service, April 23–25: 477–484.

Visualizing scenic resource impacts: proposed surface mining and solid waste sanitary landfill

Ellsworth, J.C. (2001) Design principles for recreating visual quality on surface mined landscapes, in *Post-mining land use planning and design*. Burley, J.B. (ed.), Lewiston, NY: Mellen Press.

HDR Engineering (2000) *City of Logan landfill siting study: Phase I – Landfill siting*. Prepared for Logan City Environmental Health Department. Logan, SLC: HDR Engineering.

Landscape Institute (1995) *Guidelines for landscape and visual assessment*. New York, NY: Spon Press.

Medina, A.N. (2003) Siting a proposed Cache County landfill: Visual assessment using the BLM's Visual Resource Management System, Computer Visual Simulation, and Geographic Information Systems Mapping. Unpublished Master's thesis, Department of Landscape Architecture and Environmental Planning, Utah State University, Logan.

US Department of the Interior (USDI), Bureau of Land Management, (2002a) BLM Handbook H-8400, Visual Resource Management. http://www.blm.gov/nstc/VRM/8400.html

US Department of the Interior (USDI), Bureau of Land Management, (2002b) BLM Handbook H-8400-10, Visual Resource Inventory. http://www.blm.gov/nstc/VRM/8410.html

US Department of the Interior (USDI), Bureau of Land Management, (2002c) BLM Handbook H-8431, Visual Resource Contrast Rating. http://www.blm.gov/nstc/VRM/8431.html

Watzek, K.A. and Ellsworth, J.C. (1994) Perceived scale accuracy of computer visual simulations. *Landscape Journal*, 13: 21–36.

The provision of visualization tools for engaging public and professional audiences

Appleton, K., Lovett, A., Sünnenberg, G. and Dockerty, T. (2001) Rural landscape visualization from GIS databases – a comparison of approaches, possibilities and problems. GISRUK 2001, University of Glamorgan. Proceedings (Kidner, D.B. and Higgs, G. (eds)), 195–201.

Bell, S. (1998) *Landscape: patterns, perception and process*. London: E and FN Spon, 288 pp.

Benedikt M.L. (1979) To take hold of space: isovists and isovist fields, *Environment and Planning*, B, 47–65.

Bulkeley, H. and Mol, A.O.J. (2003) Participation and environmental governance, *Environmental Values*, Special issue on Environment, Policy and Participation, Bulkeley, H. and Mol, A.P.J. (eds), Vol. 12(2), 143–155.

European Union (1998) Convention on Access to Information, Public Participation in Decision-making and Access to Justice on Environmental Matters, Aarhus, 25 June 1998.

European Union (2003) A proposal for establishing common rules for direct support schemes under the Common Agricultural Policy and support schemes for producers of certain crops, and Proposal for amending Regulation (EC) No. 1257/1999 on support for rural development from the European Agricultural Guidance and Guarantee Fund (EAGGF) and repealing Regulation (EC) No. 2826/2000, European Union, COM(2003) 23 final.

Krause, C. (2001) Our visual landscape, managing the landscape under special consideration of visual aspects, *Landscape and Urban Planning*, 54: 239–254.
Landscape Institute and Institute of Environmental Management and Assessment (LI-IEMA) (2002) *Guidelines for Landscape and Visual Impact Assessment* (2nd edn), London: Spon Press.
Miller, D. (2001) A method for estimating changes in the visibility of land cover, *Landscape and Urban Planning*, 54: 91–104.
Miller, D.R., Bell, S., Ball, J., Morrice, J.G., Wood, M., Ward-Thompson, C., Ode, Å. and Horne, P.L. (2004) Study into landscape potential for wind turbine development in East and North Highland, and Moray. A report for Scottish Natural Heritage, Macaulay Institute and Edinburgh College of Art, 198 pp.
ODPM (2003) *Participatory planning for sustainable communities: international experience in mediation, negotiation and engagement in making plans*. Wetherby: ODPM Publications.
Ordnance Survey (2003) MasterMap digital dataset of Great Britain, Ordnancesurvey.gov.uk/oswebsite/products/mastermap/
O'Sullivan, D. and Turner, A. (2001) Visibility graphs and landscaper visibility analysis, *International Journal of Geographic Information Science*, 15(2): 221–237.
Scottish Executive (2003) Scottish Planning Policy, SPP1: The Planning System, http://www.scotland.gov.uk/library5/planning/spp1-01.asp
Scottish Natural Heritage (2003) Annual Report 2002/2003, Scottish Natural Heritage, Battleby, Redgorton, Perth, 70 pp.
Steinitz, C. (1990) Toward a sustainable landscape with high visual preference and high ecological integrity in loop road in Acadia National Park, USA, *Landscape and Urban Planning*, 19: 213–250.
Tandy, C.R.V. (1967) The isovist method of landscape survey, in Murray, H.C. (ed.) *Symposium on Methods of Landscape Analysis*, Landscape Research Group, London, 9–10.

The visualization of windfarms

Benson, J.F., Jackson, S.P. and Scott, K.E. (2002) Visual assessment of windfarms: best practice. Scottish Natural Heritage Commissioned Report F01AA303A, 80 pp., Edinburgh.
Bishop, I.D. (2002) Determination of thresholds of visual impact: the case of wind turbines. *Environment and Planning B: Planning and Design*, 29(5): 707–718.
Bishop, I.D. (2003) Assessment of visual qualities, impacts and behaviours, in the landscape, using measures of visibility. *Environment and Planning B: Planning and Design*, 30: 677–688.
De Cervantes, M. (1605) *Don Quixote*, Chapter 8 (Ormsby translation, 1885).
Hankinson, M. (1999) Landscape and visual impact assessment, in Petts, J. (ed.) *Handbook of Environmental Impact Assessment*, Volume 1, Oxford: Blackwell Science, 347–373.
Kahn, R.D. (2000) Siting struggles: the unique challenge of permitting renewable energy power plants. *The Electricity Journal*, March, 21–33.
Krohn, S. and Damborg, S. (1998) On public attitudes towards wind power. *Renewable Energy*, 16 (1–4): 954–960.
Landscape Institute and Institute of Environmental Management and Assessment (LI-IEMA) (2002) *Guidelines for Landscape and Visual Impact Assessment*, (2nd edn) London: Spon Press.

Pasqualetti, M.J. (2001) Wind energy landscapes: Society and technology in the California desert. *Society and Natural Resources*, 14(8): 689–699.

Pasqualetti, M.J., Gipe, P. and Righter, R.W. (eds) (2002) *Wind power in view*. San Diego: Academic Press.

Perkins, N.H. (1992) Three questions on the use of photo-realistic simulations as real world surrogates. *Landscape and Urban Planning*, 21: 265–267.

Shang, H.-D. and Bishop, I.D. (2000) Visual thresholds for detection, recognition and visual impact in landscape settings. *Journal of Environmental Psychology*, 20: 125–140.

Stamps III, A.E. (1997) A paradigm for distinguishing significant from non-significant visual impacts: theory, implementation, case histories. *Environmental Impact Assessment Review*, 17: 249–293.

Stevenson, R. and Griffiths, S. (1994) The visual impact of windfarms: Lessons from the UK experience. ETSU W/13/00395/REP, ETSU for the Department of Trade and Industry, UK.

Thayer, R.L. Jr. (1994) *Gray World, Green Heart: Technology, Nature and the Sustainable Landscape*. New York: John Wiley and Sons Inc.

Wood, C., Dipper, B. and Jones, C. (2000) Auditing the assessment of the environmental impacts of planning projects. *Journal of Environmental Planning and Management*, 43(1): 23–47.

Wood, G. (1999) Post-development auditing of EIA predictive techniques: A spatial analytical approach. *Journal of Environmental Planning and Management*, 42(5): 671–689.

Wood, G. (2000) Is what you see what you get? Post-development auditing of methods used for predicting the zone of visual influence in EIA. *Environmental Impact Assessment Review*, 20: 537–556.

Chapter 9

Applications in the urban landscape: Introduction

Bishop, I., Lange, E. and Mahbubul, A. (2004) Estimation of the influence of view components on apartment pricing using a public survey and GIS modeling. *Environment & Planning B: Planning and Design*, 31: 439–452.

Conner, J.R., Gibbs, K.C. and Reynolds, J.E. (1973) The effects of water frontage on recreational property values. *Journal of Leisure Research*, 5(2): 26–39.

de Hollander, A.E.M. and Staatsen, B.A.M. (2003) Health, environment and quality of life: an epidemiological perspective on urban development. *Landscape and Urban Planning*, 65: 53–62.

DTLR (Department for Transport, Local Government and the Regions) (2001) Green spaces, better places. Interim report of the Urban Green Spaces Taskforce. London, 49 pp.

Duffy, B. (2003) Quality of life. *Landscape Design*, 318: 37–40.

Finke, L. (2003) Ökologische Chancen und Risiken urbaner Innenentwicklung – stadtökologische Sichtweise, in Arlt, G., Kowarik, I., Mathey, J. and Rebele, F. (eds) Urbane Innenentwicklung in Ökologie und Planung. *IÖR-Schriften*, 39: 33–48.

Gälzer, R. (2001) Grünplanung für Städte: Planung, Entwurf, Bau und Planung. Ulmer Stuttgart, 408 pp.

Hall, P. and Pfeiffer, U. (2000) *Urban future 21. A global agenda for 21st century cities*. New York: E & FN Spon, 363 pp.

Lange, E. and Schaeffer, P.V. (2001) A comment on the market value of a room with a view. *Landscape and Urban Planning*, 55: 113–120.

Rybczynski, W. (1999) *A clearing in the distance. Frederick Law Olmsted and America in the nineteenth century*. Scribner, 480 pp.

Selle, K. (1999) Impuls Landschaft. Bedeutungswandel der Freiraumfrage in der Stadtentwicklung. DISP 136/137: 36–46.

Sitte, C. (1922) *Der Städtebau nach seinen künstlerischen Grundsätzen*. Ein Beitrag zur Lösung modernster Fragen der Architektur und monumentalen Plastik unter besonderer Beziehung auf Wien (5th ed) with appendix 'Grosstadtgrün'. Wien: Graeser, 216 pp.

UNCHS (United Nations Centre for Human Settlements) (1996) *An urbanizing world. Global report on human settlements 1996*. New York and Oxford: Oxford University Press, 559 pp.

Future scenarios of peri-urban green space

Appleyard, D. (1977) Understanding professional media: issues, theory, and a research agenda, in *Human Behavior and Environment*, Altman, I. and Wohlwill, J.F. (eds) New York: Plenum Press, 43–88.

Danahy, J.W. and Hoinkes, R. (1995) Polytrim: Collaborative setting for environmental design, in Tan, M. and Teh, R. (eds) *The Global Design Studio. Proc. CAAD Futures '95, 24–26 September 1995*. CASA, National University of Singapore, 647–658.

Fachstelle für Stadtentwicklung (1999) Befragung der Einwohnerinnen und Einwohner der Stadt Zürich 1999. Erste Resultate der Befragung, 19 pp.

Fachstelle für Stadtentwicklung (2003) Befragung der Einwohnerinnen und Einwohner der Stadt Zürich 2003. Erste Ergebnisse Juni 2003, 30 pp.

Heißenhuber, A., Kantelhardt, J. and Osinski, E. (2000) Ökonomische Aspekte einer ressourcenschonenden Landnutzung, in Vorstand des Dachverbandes Agrarforschung (ed.), Entwicklung nachhaltiger Landnutzungssysteme in Agrarlandschaften. Frankfurt am Main: DLG-Verlag, *Agrarspectrum* 31: 20–30.

Hoinkes, R. and Lange, E. (1995) 3D for free. Toolkit expands visual dimensions in GIS. *GIS World*, 8(7): 54–56.

Huber, B. (1990) Die Planung der ETH-Hönggerberg 1957–1990. *DISP*, 100: 5–20.

McKechnie, G. (1977) Simulation techniques in environmental psychology. Stokols, D. (ed.) *Perspectives on Environment and Behavior*. New York NY: Plenum.

Stiens, G. (1993) Prognostische Geographie. *Geographische Rundschau*, 4: 224–231.

Stremlow, M., Iselin, G., Kienast, F., Kläy, P. and Maibach, M. (2003) Landschaft 2020 – Analysen und Trends. Grundlagen zum Leitbild des BUWAL für Natur und Landschaft. Schriftenreihe Umwelt Nr. 352, Bundesamt für Umwelt, Wald und Landschaft, Bern, 152 pp.

Swiss Federal Office for Agriculture (2003) Agricultural Report 2003, 7 pp.

Tress, B. and Tress, G. (2003) Scenario visualisation for participatory landscape planning – a study from Denmark. *Landscape and Urban Planning*, 64: 161–178.

Negotiating public view protection and high density in urban design

Allsopp, R., Able, J., Brown, T., Danahy, J.W., duToit, R., Gomes, C. and Mazur, Y. (2002) National Capital Views Protection, Final Consulting Report.

Danahy, J.W. (2001) Technology for dynamic viewing and peripheral vision in landscape visualization. *Landscape and Urban Planning*, 54: 125–138.

Danahy, J.W. and Hoinkes, R. (1995) *Polytrim: Collaborative setting for environmental design*, CAAD Futures '95, The Global Design Studio, Tam, M. and Teh, R. (eds) http://www.clr.utoronto.ca/PAPERS/CAAD95/caadf.jd8.html

Combining visualization with choice experimentation in the built environment

Adamowicz, W., Boxall, P., Williams, M. and Louviere, J. (1995) Stated preference approaches for measuring passive use values: Choice experiments and contingent valuation, Staff Paper 95-03, Department of Rural Economy, Edmonton: University of Alberta.

Blamey, R.K., Rolfe, J.C., Bennett, J.W. and Morrison, M.D. (1997) Environmental choice modelling: Issues and qualitative insights, Choice Modelling Research Report No. 4, University College, Canberra: The University of New South Wales.

Boxall, P.C., Adamowicz, W.L., Swait, J., Williams, M. and Louviere, J. (1996) A comparison of stated preference methods for environmental valuation, *Ecological Economics*, 18: 243–253.

Chadwick, A. (2002) Socio-economic impacts: Are they still the poor relations in UK environmental statements?, *Journal of Environmental Planning and Management*, 45: 3–24.

Davies, A. and Laing, R. (2002a) Designing choice experiments using focus groups: Results from an Aberdeen case study, *Forum Qualitative Sozialforschung/Forum: Qualitative Social Research*, Online Journal, 3(3), available at: http://www.qualitative-research.net/fqs/fqs-eng.htm

Davies, A. and Laing, R.A. (2002b) Streetscapes: Their contribution to wealth creation and quality of life, final report to Scottish Enterprise, Aberdeen: The Robert Gordon University.

Davies, A. and Laing, R. (2003) Images and stated preferences: Do people need to be told what the attributes are or do they notice them anyway?, in Craig, A. (ed.) *Proceedings of EPUK '03*, Aberdeen: The Robert Gordon University.

Hanley, N., MacMillan, D.C., Wright, R.E., Bullock, C.H., Simpson, I., Parsisson, D. and Crabtree, R. (1998) Contingent valuation versus choice experiments: Estimating the benefits of Environmentally Sensitive Areas in Scotland', *Journal of Agricultural Economics*, 49: 1–15.

Kahneman, D. and Knetsch, J.L. (1992) Valuing public goods: The purchase of moral satisfaction, *Journal of Environmental Economics and Management*, 22: 57–70.

Morrison, M.D., Blamey, R.K., Bennett, J.W. and Louviere, J.J. (1996) A comparison of stated preference techniques for estimating environmental values, Choice Modelling Research Report No. 1, University College, Canberra: The University of New South Wales.

Chapter 10

Visualization prospects – Introduction

Azuma, R., Baillot, Y., Reinhold, B., Feiner, S., Julier, S. and MacIntyre, B. (2001) Recent advances in augmented reality. *IEEE Computer Graphics and Applications*, 21: 34–47.

Bureau of Land Management (1980) Visual simulation techniques, US Deptartment of the Interior, Washington, D.C.: US Government Printing Office (Stock no. 024-001-00116-6).

Jepson, W.S., Liggett, R.S. and Friedman, S. (2001) An integrated environment for urban simulation. *Planning support systems: Integrating geographic information systems, models and visualization tools*. Brail, R.K. and Klosterman, R.E. (eds), Redlands, CA: ESRI Press, 387–404.

Perrin, L., Beauvais, N. and Puppo, M. (2001) Procedural landscape modeling with geographic information: the IMAGIS approach. *Landscape and Urban Planning*, 54: 33–48.

van Maran, G. and Germs, R. (1999) Karma VI: a virtual reality interface for the spatial data engine. ESRI User Conference Proceedings (gis.esri.com/library/userconf/proc99/navigate/proceed.htm)

Verbree, E., van Maren, G., Germs, R., Jansen, F. and Kraak, M.-J. (1999) Interaction in virtual world views – linking 3D GIS with VR. *International Journal of Geographical Information Science*, 13: 385–396.

Compositing of computer graphics with landscape video sequences

Debevec, P. (1998) Rendering synthetic objects into real scenes, *Proc. SIGGRAPH '98*: 189–198

Li, M. and Lavest, J. (1996) Some aspects of zoom lens camera calibration, *IEEE Trams. Pattern Analysis and Machine Intelligence*, 18: 1105–1110.

Nakamae, E. (2001) An overview of photo-realism for outdoor scenes, *International Journal of Image and Graphics*, 1: 27–43.

Nakamae, E., Harada, K., Ishizaki, T. and Nishita, T. (1986) A montage method: the overlaying of the computer generated images onto a background photograph, *Proc. SIGGRAPH '86*, 13: 207–214.

Nakamae, E., Qin, X. and Tadamura, K. (2001) Rendering of landscapes for environmental assessment, *Landscape and Urban Planning*, 54: 19–32.

Nakamae, E., Qin, X., Jiao, G., Rokita, P., Tadamura, T. and Usagawa, Y. (1998) Computer generated still images composited with panned landscape video sequences, *Proc. MultiMedia Modeling '98 (MMM98)*, 61–70.

Nishita, T., Dobashi, Y. and Nakamae, E. (1996) Display of clouds taking into account multiple anisotropic scattering and skylight, *Proc. SIGGRAPH '96*: 379–386.

Qin, X., Nakamae, E. and Tadamura, K. (2002) Automatically compositing still images and landscape video sequences, *IEEE Computer Graphics and Applications*, 22: 68–78.

Qin, X., Nakamae, E. and Tadamura, K. (1999) Creating a precise panorama from panned video sequence images, *Proc. Pacific Graphics '99*, 6–11.

Qin, X., Nakamae, E., Tadamura, K. and Nagai, Y. (2003) Fast photo-realistic rendering of trees in daylight, *Computer Graphics Forum (Proc Eurographics 2003)*, 22: 243–252.

Shum, H.Y. and Szeliski, R. (1998) Construction and refinement of panoramic mosaics with global and local alignment, *Proc. Sixth IEEE International Conference On Computer Version*, 953–958.

Szeliski, R. (1996) Video mosaics for virtual environments, *IEEE Computer Graphics and Application*, 16: 251–258.

Tadamura, K., Qin, X., Jiao, G. and Nakamae, E. (1999) Rendering optimal solar shadow using plural sunlight depth buffers, *Proc. Computer Graphics International '99*, 166–173.

Future use of augmented reality for environmental and landscape planners

Azuma, R. (1997) A survey of augmented reality. *Presence: Teleoperators and virtual environments*, Vol. 6, No. 4, 355–385.

Azuma, R., Baillot, Y., Behringer, R., Feiner, S., Julier, S. and MacIntyre, B. (2001) Recent advances in augmented reality. *IEEE Computer Graphics and Applications*, Vol. 21, No. 6, 34–47.

Azuma, R.T. (1997) The challenge of making augmented reality work outdoors, in *Mixed reality: Merging real and virtual worlds*, 379–390, March 1999.

Bolt, R.A. (1980) 'Put-That-There': Voice and gesture at the graphics interface, in *ACM SIGGRAPH '80*, Seattle, WA, July 1980, 262–270.

Brooks, F.P. (1988) Grasping reality through illusion – interactive graphics serving science, in *Computer–Human Interaction CHI 88*, Washington, DC, May 1988, 1–11.

Buxton, W. and Myers, B.A. (1986) A study in two-handed input, in *CHI – Human Factors in Computing Systems*, Boston, MA, 1986, 321–326.

Hinckley, K., Pausch, R., Goble, J.C. and Kassell, N.F. (1994) A survey of design issues in spatial input, in *7th International Symposium on User Interface Software Technology*, Marina del Rey, CA, November 1994, 213–222.

Piekarski, W. and Thomas, B.H. (2001) Tinmith-Metro: New outdoor techniques for creating city models with an augmented reality wearable computer, in *5th International Symposium on Wearable Computers*, Zürich, Switzerland, October 2001, 31–38.

Piekarski, W. and Thomas, B.H. (2003) An object-oriented software architecture for 3D mixed reality applications, in *2nd International Symposium on Mixed and Augmented Reality*, Tokyo, Japan, October 2003.

Visualization and participatory decision making

Arnstein, S. (1969) A ladder of citizen participation. *Journal of the American Institute of Planners*, 35: 45–54.

Perkins, N.H. and Barnhart, S. (1996) *An assessment of visual simulations used to demonstrate the visual impact of development within the Rouge River valley corridor*. Technical report prepared for the Ontario Municipal Board.

Pretty, J. (1995) Participatory learning for sustainable agriculture. *World Development* 23, 8: 1247–1263.

Repton, H. and. Repton, J.A. (1816) Fragments on the theory and practice of landscape gardening. London: T. Bensley and Son.

Rocha, E. (1997) The ladder of empowerment. *Journal of Planning Education and Research*, 17: 31–44.

Sheppard, S. (1989) *Visual simulation: A user's guide for architects, engineers, and planners*. New York: Van Nostrand Reinhold.

Smardon, R., Palmer, J. and Felleman, J. (eds) (1986) *Foundations for visual project analysis*. New York: John Wiley & Sons.

Visualization in support of public participation

Appleyard, D. (1979) The environment as a social symbol: within a theory of environmental action and perception, *Journal of American Institute of Planners*, 45: 143–152.

Baltimore Regional Transportation Board (2003) Vision 2030: Shaping our Region's Future Together (http://www.baltometro.org/pdfs/V2030report.pdf).

Trends, challenges and a glimpse of the future

Feeney, M-E.F., Williamson, I.P. and Bishop, I.D. (2002) The role of institutional mechanisms in spatial data infrastructure development that supports decision making, *Cartography*, 32: 21–37.

Kourepenis, A., Borenstein, J., Connelly, J., Elliot, R., Ward, P. and Weinberg, M. (1998) *Performance of MEMS inertial sensors*. IEEE Position Location and Navigation, Palm Springs, CA, 1–8.

OpenGIS (1998) The OpenGIS Guide: Introduction to interoperable geoprocessing and the OpenGIS specification, 3rd edn, accessed December 2004 at www.geo.ulg.ac.be/interne/Fichiers_Interet_General/SIG/OpenGISGuide.pdf

INDEX

3D StudioViz 169
3D-MapEditor 158
3D-Player 158, 159

Aarhus Declaration 175
abstraction 28, 29
active stereo 73
aesthetic impacts 166
  data reliability, accuracy, currency issues 171
  landfill 168–70
  legal/procedural issues 170–171
  political/social factors 171–172
  simulation issues/process 170–173
  surface mining 166–168
  visualization process 172–173
agriculture, assessment of visualized landscapes 141–144
  background 133–135
  biodiversity conservation 139
  business as usual 138
  initial fieldwork/scenarios 138–139
  landscape character 138
  positive response of farmers 143–144
  research context/objectives 136–138
  supplemented biodiversity conservation 139
  visualization of landscape scenarios 139–141, *see also* rural communities
animals 17–18, 196–201
animation 8–9, 69, 75

ArcGIS 146–147
Arrow Innovative Forest Practices Agreement (IFPA) project 121
  application 123–124
  implications 124
  technology development 121–122
arterial route selection 245–246
atmosphere and light 19–20
augmented reality (AR) 14–15, 223–224, 264
  background 234
  building design 239
  future possibilities 240
  landscape design 236–238
  placement of predefined graphical objects 236–238
  specification of planar shapes 238
  technology 235–236
Australia 112–119, 145–151
Autodesk 169

Baltimore case study 252–259
Barker, Robert 6–7
Barrick-Mercur Gold Mine 166–168
bathymetry 44
Berkeley Environmental Simulation Laboratory 8
Boundary Waters Canoe Area Wilderness 107
Breysig, Adam 7
British Columbia 89, 91, 94, 121
Brunelleschi, Filippo 6

buildings, data 45–46
  parametric representation 45–46
  procedural geometry 45–46
built structures 18–19
  visual impact 198–201, 205–208
Buscot and Coleshill Estate (Oxfordshire) 136

Cache County Landfill Siting Study 168
calibrated image 104
  close in scale/time 104
  distant in scale/time 104–106
  middle-ground challenge 106
Cambridge (Ontario) 245–246
Catal Hüyük (Turkey) 5
cave paintings 4
Central Park (New York) 193
Centre for Landscape Research (CLR) 203, 204, 208, 210
choice experiments 212–213, 216
  attribute-relate results 218
  experimental work 216–217
  focus groups 213–214
  image realism/related issues 217–218
  technical limitations 219
  validity issues 218–219
  visualization 213, 219
Collaborative for Advanced landscape Planning (CALP) 89, 120–124
Common Agricultural Policy (CAP) 176

communication 25
  between landscape professionals 179–182
  interactive computer visualization 178
  and landscape change 176
  with members of the public 176–179
  and photomontage 179
  physical model 177, 178–179
  professionals 179–182
  public 176–179
CommunityViz 89, 95, 253
computer graphics 10
computer power, frame-rate 52, 53–54
  graphics card specifications 54–55
  performance 52–55
  rendering 51–52
  significance 51–55
computer-aided design (CAD) 8, 19, 45, 62, 66, 132, 171, 196, 208, 264
computer-generated images 228–231
coordinate reference systems (CRS) 36–38
  geographic coordinates 36–37
  geographical datum 37
  Transverse Mercator (TM) 37
  Universal Transverse Mercator (UTM) 37
Countryside Agency 136

data sources 48–49
  basic concepts 35–36
  buildings/artificial structures 45–46
  CRS 36–8
  culture 35
  elevation 38–40
  image 40–43
  paper maps 36
  transportation 46–48
  vegetation 43–44
  water 445
Defense Mapping Agency (USA) 8
Department for Transport, Local Goverment and Regions (DTLR) (UK) 193
desktop graphics 261
dimensionality 27–28
dynamic representation 26

elevation data, breakline 40
  DEM format 38

grid formats 39
hypsography 40
triangular irregular networks (TIN) 38, 40
Emerald Necklace (Boston) 193
English Nature 136
environment 29
Environment Analysis (EA) 170, 172
Environmental Impact Assessment (EIA) 184–185, 190
Environmental Impact Statement (EIS) 170, 172
Environmental Visualisation 29
Envisioning System (EvS) 145–146
ERDAS IMAGINE VGIS software 178
ethics 80–81
  design for new crystal ball 92–96
  interim code 86–88
European Petroleum Survey Group (EPSG) 38

Finland, forestry visualization techniques 125–132
Flash 217
Florence (Italy) 6
forestry 90
  alternative practices 112
  animation/display 117–118
  applications of systems in practice 131–132
  background 101–104
  calibrated images 104–106, 110–111
  case study 107–109
  composition/condition 105
  data 44
  data problems 106
  determination of species mix 114
  development of individual trees 115
  diameter at breast height (DBH) 44
  distribution of mix over terrain 115
  field evaluation 119
  financial aspects 125
  Finnish perspective 125–132
  harvesting 112
  inventory/descriptions 105
  masking of object 110
  modelling/digital representation 120–124
  preliminary simulation/assessment 113
  public response to harvest systems 118

realistic simulation process 113–114
review of simulators 127–131
scene description file/rendering 116–117
spatial location 110
visualization tools 107–109, 112–118, 121–124, 125–132
wood procurement 125
Forestry Tasmania 112, 115
FORSI 127-8

Ga-Sur (Yorghan Tepe) (Iraq) 5
games software/technology, development 62–65
  graphics processing unit (GPU) 64–65
  landscape planning/design process 65–66
GeoCommunity 36
geographic information system (GIS) 9, 11, 20, 27, 29, 88, 163
  agriculture 145, 147–151
  buildings 45
  forestry 44
  Lenné3D 157
  three-dimensional linkage 224–225
  urban space 196
georeferenced 35
geospecific 31, 35, 42
geotypical 31, 35, 42
Giotto 6
Global Environment Change 136
global positioning system (GPS) 262
green space 193–194
  Aberdeen 212–219
  Ottawa/Toronto 203–210
  Zürich 195–201
Gunflint Trail 107–109

Heilig, Morton 12
Hudson River 163
humans 17–18
hydrography 44

Ikonos 16
image data, aerial 41
  bands 40
  file formats 41–42
  multispectral 40
  panchromatic 40
  satellite 41
  surface texture 42–43
  TIFF files 43

Index

image editing, digital photomontage 9, 164
  image processing 104–105, 134, 166–170, 208
  texture extraction 115
IMAGIS 224
immersive displays, CAVE 72–3, 232–233
  future of 263
  head-mounted display 72, 235–236
  multi-screen immersion 72–73
  use of 117, 151, 208, 239
  value of 95–96, 219
infrastructure 163–165
Institute of Environmental Management Assessment 179
integrated systems 225
interaction 27, 70, 93–96
Interim Code of Ethics 86–88
Internet 12, 74–75, 92, 216–217, 244–245

Käferberg case study 195–196
  implications for landscape planning 201
  objectives 196
  results 199–201
  scenarios 197–198
  survey 198
  visualization methodology 196–197
knowledge 25
kriging 36

L-System 17
Laboratory for Computer Graphics and Spatial Analysis (Harvard Graduate School of Design) 9
Land Management Initiatives 136
LandEx 158
landscape 6, 10
  agricultural 138, 139–144
  change 176
  design 236–238
  early visualization techniques 6, 10
  perceptual/societal issues 20–21
  planning 65–66, 201
  synthetic 57–60
  types 15–20
  visualized 141–144
  walking 157, *see also* urban landscape
Landscape Institute 179
Lanier, Jaron 12
laser scanning 19

Lenné, Peter Joseph 153
Lenné3D 152
  aims/challenges 153, 161–162
  beginning 155
  editing in 3D 157
  first person view 157
  initial state 155–156
  involvement 153, 155
  planning with GIS 157
  plant distribution 156
  plant modelling 156–157
  reaction to 160–161
  walk through 153–157
Lifescapes project 136

maps 4–5, 36
MicroElectro Mechanical Systems (MEMS) 262
Minnesota 107
mixed reality (MR) 15
modelling, development 56–57
  problems 61
  rendering terrain 60
  synthetic landscapes 57–60
MONSU 128–130, 131
multi-sensory presentations 70–71
Murten panorama 7

National Trust (UK) 136, 143
Nippur 5
Nouvel, Jean 7

Olmsted, Frederick Law 6, 193
on-line surveys 109, 198, 216
Ontario 107
OpenGIS Consortium (OGC) 38, 262
OpenGL 19, 95, 146
OpenGL Performer 146–147
orientation 261–262
Ottawa-Gatineau 205–207

Palenque 75
panoramas 6–7, 68–69
panoramic images 226
  compositing from video sequence frames 226
  making background image 227
passive stereo 73
peri-urban green space 195–201
photo-realistic computer virtual simulations 174
photography 4, 7, 16
photomontage 9, 179, 186–188, 189

planned landscapes 152
  walk-through visualization technique 153–162
Planning for Real 178
polygon-based approach 17
Polytrim 196, 203
positioning 261–262
presentation, animations 69
  human factors 76–77
  interactive visualizations 70
  multi-screen immersion 72–73
  multi-sensory presentation 70–71
  panoramas 68–69
  printing 71
  single screen 71–72
  stereo display 73–75
  World Wide Web 74–75
printing 71
public participation 223–234
  background 251–252
  case studies 138–144, 195–196, 245–250, 252–259
  community values 145–151
  development patterns 255
  participatory decision-making 241
  public engagement 176–179
  public meetings 253–259
  scenario comparisons 256–258
  Thematic Committees 255, 258–259
  visioning (envisioning) 145–6, 252–253

QuickBird 16

radiosity 20
raster to vector conversion 36
ray tracing 20
real-time kinetics (RTK) 262
real-time visualization 145–151, 152–161, 203–211
realism 28–31
Red Hill Creek (Hamilton, Ontario) 246–250
reliability 84
Repton, Humphrey 6, 242
River Cole 136
River Thames 136–142
roads, carving 47
  draping 47
  merging 47
rural communities, envisioning systems 145–146
  impact models 150
  land cover manipulation 147–148

rural communities, envisioning systems (*continued*)
  navigation 148–149
  outlook 151
  simulation environment 146–147
  using interaction/feedback 147
  voting 149, *see also* agriculture

Sahara 4
scientific visualisation 10
Scotland 185–191
Scott Sutherland School (Robert Gordon University, Aberdeen) 212–219
Scottish Natural Heritage (SNH) 19, 175
Sitte, Camillo 193
SmartForest 106, 107, 130–131
solid waste sanitary landfill project 168–170
Spatial Data Authority 263
spatial data infrastructure (SDI) 36, 261
spatial reference system (SRS) 38
Stand Visualization System (SVS) 106
stereo display 73–74
Stockholm Earth Summit (1972) 163
surface mining 166–168
Sutherland, Ivan 12

terrain 15–16
  rendering 60
texture map 17, 42
Thun (Switzerland) 7
Tinmith-Endeavour 236–240
Toronto project 208–210
transport, data 46–48
  levels of detail 46
  rendering of roadways 47

University of British Columbia (UBC) 105, 120
University of Toronto 203
urban landscape, background 193–194
  peri-urban green space 195–201

validity, accuracy 83
  actual/apparent realism 82–83
  construct 82
  content/face 81
  criterion-related/predictive 81–82
  dimensions 82–84
  visual clarity 83

vegetation 17
  billboards 44
  data 43–44
  directional billboards 44
  explicit modelling 44
video sequences 224
  automatic camera tracking 229–231
  basic geometrical relationship 229
  CAVEs 232–233
  compositing panoramic image 226
  frames 231
  future problems 232–233
  moving camera/computer-generated images 231
  tripod/computer-generated still images 228–229
virtual environment 32–4
  binary tree 67
  creating 66–67
  potential visibility set (PVS) 67
Virtual Reality Modelling Language (VRML) 12, 129–130, 139–144, 217
virtual reality (VR) 12–15, 32–34, 145, 148, 248
Vision 2030: Shaping our Region's Future Together (Baltimore) 252–259
Visual Impact Assessment (VIA) 190, 191
Visual Nature Studio 88
Visual Resource Management (VRM) System 168
visual simulation (Vis-sim) 35
visualization, adoption of term 23
  communicating vs discovering knowledge 25
  definition 23–24
  history of 4–12
  image validity/manipulation 10–11
  importance 3–4
  levels of interactivity 27
  levels of realism 28–31
  mapping 4–5
  perspective 5–6
  principles 85–86
  public demand 263
  quality thresholds 85
  realism 186–188
  role of 28–29
  scientific 10
  single vs multiple representations 32
  spatial/temporal 26
  static/dynamic display 26
  terrain representation 5

three-dimensional representation 5–12
trends, challenges, future prospects 261–265
variables 24
visualization in scientific computing (ViSC) 76–77
visualization systems, crystal ball 79–81, 88–96
  issues 79
  outcomes 80
  support infrastructure 80
  terminology 80
visualization tools 80
  arguments against needing better crystal ball 88–89
  arguments for needing better crystal ball 89–92
  engagement with public 175–183
  ethical design for new crystal ball 92–96
  ethics 80–81
  flexible/interactive approaches 93–96
  forestry 107–9, 112–118, 121–124, 125–132
  limitations 90
  prescriptive approaches 92–93
  progressive refinement, comparison, negotiation 204

Warra Long Term Ecological Research (LTER) 112
water 18
  data 44–45
  sample points 44–45
wind turbines/windfarms 164–165, 176
  accuracy in environmental impact assessment 185–186
  development process 184–185
  dynamic technologies 188–190
  model walk-through 179–182
  opposition to 184
  perception/significance 190–191
  pro-lobby/anti-lobby debates 185
  realism in visualization/photomontage 186–188
Wocher, Marquard 7
World Construction Set (WCS) 88, 89, 91
World Wide Web *see* Internet

Xfrog 57–58, 157, 159

Zones of Visual Influence (ZVI) 179, 185–186, 188
Zürich (Switzerland) 195–201